Sensory-Directed Flavor Analysis

T0132569

FOOD SCIENCE AND TECHNOLOGY

Editorial Advisory Board

Sensory-Directed Flavor Analysis

Edited by

Ray Marsili

CRC Press
Taylor & Francis Group
Boca Raton London New York

CRC Press is an imprint of the
Taylor & Francis Group, an **informa** business
A TAYLOR & FRANCIS BOOK

CRC Press
Taylor & Francis Group
6000 Broken Sound Parkway NW, Suite 300
Boca Raton, FL 33487-2742

First issued in paperback 2019

ISBN-13: 978-1-57444-568-8 (hbk)
ISBN-13: 978-0-367-39039-6 (pbk)

Library of Congress Cataloging-in-Publication Data

Sensory-directed flavor analysis / edited by Ray Marsili.
    p. cm. -- (Food science and technology ; 163)
  Includes bibliographical references and index.
  ISBN 1-57444-568-5 (alk. paper)
  1. Food--Sensory evaluation. 2. Flavor. I. Marsili, Ray, 1946- II. Series: Food science and technology (Taylor & Francis) ; 163.

TX546.S454 2006
664'.072--dc22
                                               2006045647

Visit the Taylor & Francis Web site at
http://www.taylorandfrancis.com

and the CRC Press Web site at
http://www.crcpress.com

Dedication

To my grandchildren: Jacob, Hillary, and Charlie

… If you are to have but one book with you upon a journey, let it be a book of science. When you read through a book of entertainment, you know it, and it can do no more for you; but a book of science is inexhaustible.

James Boswell (1740–95)

Scottish author, biographer of Samuel Johnson

Preface

Aroma and flavor are primary contributors to the appeal of foods and beverages and, as a consequence, are critically important to food industry researchers, new product developers, product formulators, and manufacturers. Today, flavor chemists have extraordinary sample preparation tools, sophisticated GC-MS instruments and time-saving automation accessories at their disposal for studying flavor problems. For example, with automated solid-phase microextraction (SPME) methods followed by GC-MS analysis, flavor chemists can generate copious amounts of chemical data in a short time with relatively little effort. However, more data does not necessarily translate into more understanding. Commonly reoccurring questions continue to confound flavor chemists: How do we determine which chemicals are impacting flavor perception? How do we accurately and reproducibly measure flavor perception based on volatile chemical profiles? How do we examine 100 or more GC peaks and decide which peaks are important to the flavor problem we are investigating? Indeed, how do we even know if we have extracted the appropriate flavor chemicals?

Our ability to extract, isolate, and concentrate potential flavor-important chemicals from complex food systems has surpassed our ability to understand how the chemical data relates to flavor. Without a systematic, scientific approach, the revelations buried in the countless number of chromatographic peaks may be misinterpreted or overlooked completely. This book is an attempt to help flavor chemists unlock the flavor secrets that may be hiding in their chromatograms; it attempts to help flavor chemists translate cold, hard numbers into a better understanding of the sense of smell and taste.

Sensory scientists employ trained taste panels to generate objective sensory data and sophisticated statistical software to help them interpret the sensory significance of their data, whereas analytical chemists apply increasingly sensitive instrumental techniques to resolve their flavor issues. When it comes to solving problems, each approach has its advantages and disadvantages, depending on the nature of the problem being studied and the specific flavor chemicals involved. Some of the examples presented in this book show sensory analysis is the most efficient problem-solving tool; other examples illustrate chemical analysis is the preferred approach, but most examples show that an approach incorporating both sensory and chemical analysis is the optimum problem-solving strategy.

One objective of *Sensory-Directed Flavor Analysis* is to encourage sensory scientists to incorporate more analytical data along with sensory data in their studies and to encourage analytical flavor chemists to include more sensory techniques in understanding their problems. Another objective of this book is to show flavor chemists how to apply ancillary techniques to help elucidate how various chemical constituents are influencing food flavor and appeal. The important enabling-technologies discussed include GC-olfactometry (GC-O) and the combination of GC-O with multidimensional GC, as well as the application of odor activity values (OAVs), multivariate statistical analysis, model system studies (a.k.a., recombination studies), etc. Additional chapters discuss analytical techniques such as solid-phase dynamic extraction and preseparation techniques in aroma analysis. The final chapter on character-impact flavor compounds is an expanded, updated version of a similar, highly-popular chapter that appeared in *Flavor, Fragrance and Odor Analysis* (Marcel Dekker, 2002); it is included at the end of the book so it is easy to find; flavor scientists will likely refer to it often in their daily work. Included throughout are a broad array of applications with dozens of tables, graphs, chromatograms, and figures.

The Editor

Ray Marsili has published over 30 papers on food and flavor chemistry in refereed chemistry journals, as well as over 100 technical papers for trade publications and numerous chapters for flavor chemistry and dairy chemistry books. He served as a contributing editor of *Food Product Design* magazine, editor of *R&D* magazine's *Sample Preparation Newsletter*, and lecturer in organic chemistry at Rock Valley College, Rockford, Illinois. He has edited two previous books on flavor and odor analysis: *Techniques for Analyzing Food Aroma* (1997) and *Flavor, Fragrance and Odor Analysis* (2002). He has over 30-years' experience as an R&D analytical laboratory manager/flavor chemist for major national and international food and ingredient companies. Today, he is an independent analytical flavor consultant, with a laboratory at Rockford College, Rockford.

Acknowledgments

I extend special appreciation to this book's contributors — many of whom have previously contributed to other flavor chemistry books I have edited. I would like to thank them not only for their outstanding professional contributions to this book and to the field of flavor chemistry but also for their friendship and support, as well as all they have taught me over the years. I would also like to thank Professor Fred Hadley and Rockford College for providing me a laboratory and office in which to continue my flavor research studies. Last but not least, I thank my wife Deborah for her support and patience.

Ray Marsili
Rockford, Illinois

Contributors

Helen M. Burbank
Department of Food Science
and Technology
Oregon State University
Corvallis, Oregon

K.R. Cadwallader
Department of Food Science
and Human Nutrition
University of Illinois
Urbana, Illinois

A.D. Caudle
Department of Food Science
Southeast Dairy Foods Research
Center
North Carolina State University
Raleigh, North Carolina

Ingo Christ
Chromsys LLC
Alexandria, Virginia

M.A. Drake
Department of Food Science
Southeast Dairy Foods Research
Center
North Carolina State University
Raleigh, North Carolina

David K. Eaton
Microanalytics
Round Rock, Texas

Y. Fu
Kerry Ingredients
Brookfield, Wisconsin

S. Karow
Kerry, Inc.
Beloit, Wisconsin

Ulrike B. Kuehn
Chromsys LLC
Alexandria, Virginia

T. Laban
Kerry Ingredients
New Century, Kansas

Ray T. Marsili
Marsili Consulting Group
Rockford College
Rockford, Illinois

Robert J. McGorrin
Department of Food Science
and Technology
Oregon State University
Corvallis, Oregon

R.E. Miracle
Department of Food Science
Southeast Dairy Foods Research
Center
North Carolina State University
Raleigh, North Carolina

Lawrence T. Nielsen
Microanalytics
Round Rock, Texas

Michael C. Qian
Department of Food Science
  and Technology
Oregon State University
Corvallis, Oregon

Yuanyuan Wang
Department of Food Science
  and Technology
Oregon State University
Corvallis, Oregon

Ken Strassburger
Degussa Fruit Flavors
Cincinnati, Ohio

Donald W. Wright
Microanalytics
Round Rock, Texas

Table of Contents

chapter one

Comparing sensory and analytical chemistry flavor analysis

Ray T. Marsili

Contents

One of the earliest users of sensory analysis was the dairy industry. In the early 1900s, techniques for judging dairy products were developed to stimulate interest and educate people in dairy science. Judging and grading dairy products normally involve assigning quality scores to products by one or two trained "experts," and evaluations are not replicated. Attributes scored include appearance, flavor, and texture, based on the presence or absence of predetermined defects. This approach has provided the dairy industry with a body of knowledge on sensory defects and their causes, and although these traditional methods are valuable for rapid product quality assessment in a hectic industrial setting, they are, in general, not useful for product innovation and development of new products that meet consumer acceptance.

Modern sensory analysis has evolved into an extremely useful tool for flavor researchers. One thing in common to all sensory assessment methods is that they use human subjects as the measuring instrument. In general, there are two major types of sensory analysis: affective and analytical. Affective sensory tests are based on consumers and their perceptions of acceptability, and are important to the food industry because they explain the role of flavor, texture, and appearance in influencing consumer acceptability. These types of techniques can only measure what untrained consumers think; they tend to suffer from extensive person-to-person variability. Therefore, polling a large number of consumers (> 50) is typically done to improve the statistical validity of the information obtained.

The second type of sensory methods, analytical techniques, is based on trained panelists. These include discriminatory tests (difference and threshold), as well as descriptive sensory analysis, perhaps the most powerful of all sensory tools. This technique is well suited for both identifying flavors in a product and discriminating sensory properties between products.

Popular difference tests include triangle tests — the panel attempts to detect which one of three samples is different from the other two — and duo–trio tests — the panel selects which one of two samples is different from a standard. One deficiency of these tests is that the nature of the differences is not defined. Although difference tests are useful for guiding product develop efforts, they should not be relied on too heavily.

Descriptive sensory analysis refers to a collection of techniques that seek to discriminate between a range of products based on their sensory characteristics and to determine a quantitative description of the sensory differences that can be identified, not just the defects. Unlike traditional quality judging methods that have been used by dairy scientists, no judgment of "good" or "bad" is made because this is not the purpose of the evaluation.

Descriptive sensory analysis methods provide useful information for flavor research, product development, and food product marketing that is lacking in traditional quality judging methods.

I. Quantitative descriptive analysis

The first published descriptive sensory technique is the flavor profile method (FPM), developed in the 1950s by Arthur D. Little, Inc. Refinements and variations in FPM occurred in the 1970s with the development of quantitative descriptive analysis (QDA) and the Spectrum™ method of descriptive analysis.

Today, QDA has gained wide acceptance as one of the most important tools to study problems related to flavor, appearance, and texture, as well as a way to guide product development efforts. For example, it has been used as an investigative sensory technique for studying conventionally pasteurized milk [1,2], ice cream [3,4], and cheese [5].

With QDA, selected panelists work together in a focus group to identify key product attributes and appropriate intensity scales specific to the product under study. The panelists are then trained by the panel leader, a sensory professional rather than a member of the panel, to reliably identify and score product attributes. During training, the panel (usually 8 to 12 prescreened individuals) generates the language (or lexicon) to describe the product. The trained panelists generate the attribute terms and descriptors that are meaningful to consumers. Therefore, the QDA profile information can be applied to modeling predictions of consumer acceptability. QDA results are subjected to statistical analysis and are then represented in a variety of graphical formats for interpretation.

One useful statistical technique is principal component analysis (PCA), a multivariate analysis method that can be used to show groupings or clusters of similar sample types based on quantitative measurements of sensory attributes. By applying PCA to QDA data, the set of dependent variables (i.e., attributes) is reduced to a smaller set of underlying variables (called factors) based on patterns of correlation among the original variables [6]. The factors (also called principal components) are linear combinations of the independent variables.

The resulting data can then be applied in many useful ways. A few examples include profiling specific product characteristics, comparing and contrasting similar products based on attributes important to consumers, and altering product characteristics with the goal of increasing market share for a given set of products.

II. Flavor lexicons

Development of an appropriate flavor lexicon is critical to the success of QDA. M.A. Drake and G.V. Civille [7] have reviewed lexicon history, methods, and applications. A flavor lexicon is a set of word descriptors that describe a product's flavor. Although the panel generates its own list to describe the product array under study, a lexicon provides a source of possible terms with references and definitions for clarification.

According to Drake and Civille, development of a representative flavor lexicon requires several steps, including appropriate product frame-of-reference collection, language generation, and designation of definitions and references, before a final descriptor list can be determined. Once developed, flavor lexicons can be used to record and define product flavor, compare products, and determine storage stability, as well as interface with consumer liking, acceptability, and chemical flavor data.

Good flavor lexicons should be both discriminating and descriptive. The language should be developed from a broad representative sample set that exhibits all the potential variability within the product. This could involve a considerable effort. For example, Drake et al. [8] collected 220 samples of cheddar cheese varying in age, milk heat treatment, and geographical origin to identify a descriptive language for cheddar cheese (see Table 1.1). Seventy samples were screened in order to develop the cheddar cheese lexicon.

In creating a lexicon, the panel will frequently review the list, merging like terms, eliminating redundancies, and organizing the list so that the attributes appear in most products being tested. It is important that multiple terms not be used to describe the same flavor; conversely, it is also important that one term does not represent or overlap with several other flavors. As an example of this type of lexicon problem, Drake et al. [8] reported that use of the term "aged" in a cheddar cheese flavor lexicon was in fact a meta term that comprised three flavors and one basic taste.

An optimized lexicon has the ability to relate consumer acceptance or rejection and instrumental or physical measurements; it can readily relate to both consumer and instrumental measurements.

The use of chemical components (sometimes referred to as chemical anchors), particularly those isolated from the product under study, can make a lexicon more understandable, unambiguous, and grounded, establishing a link to formulation and manufacture of that product. Creating this type of link can be time-consuming and challenging. However, even without chemical references, a lexicon can be discriminative and precise.

Dairy flavor researchers have done an excellent job, applying flavor lexicons and QDA to the resolution of flavor problems. Several different flavor lexicons have been developed to study aroma and flavor development in dairy products, the effects of fat reduction, and the effects of different starter or adjunct bacteria. Muir et al. [9] described nine aroma terms for characterization of aroma profiles of hard and semihard cheese. For studying flavor development in cheddar cheese during maturation, Piggot and Mowat [10] determined 23 descriptive flavor terms, and Roberts and Vickers [11] developed a flavor lexicon. Muir et al. [12] and Drake et al. [13,14] used descriptive sensory panels to determine the effect of starter culture and adjunct cultures on cheddar cheese flavor. Banks et al. [15] used QDA to determine sensory properties of low-fat cheddar cheese.

Table 1.1 Cheddar Cheese Lexicon with Identified References

Term	Definition	Reference
Cooked	Aromatics associated with cooked milk	Nonfat milk heated to 85°C for 30 min
Whey	Aromatics associated with cheddar cheese whey	Fresh cheddar whey
Diacetyl	Aromatics associated with diacetyl	Diacetyl
Milk fat/ lactone	Aromatics associated with milk fat	Fresh coconut meat, heavy cream, δ-dodecalactone
Fruity	Aromatics associated with different fruits	Fresh pineapple, canned pineapple juice
Sulfur	Aromatics associated with sulfurous compounds through water, struck match	Boiled mashed egg, H_2S bubbled
Free fatty acid	Aromatics associated with short-chain fatty acids	Butyric acid
Brothy	Aromatics associated with boiled meat or vegetable soup stock	Canned potatoes, commercial low-sodium beef broth cubes, vegetable broth cubes, beef broth cubes
Nutty	Nutlike aromatic associated with different nuts	Lightly toasted unsalted nuts, wheat germ, unsalted wheat crackers
Catty	Aromatics associated with tomcat urine	2 Mercapto-2-methyl-pentan-4-one
Cowy/ phenolic	Aromatics associated with barns and stock trailers	p-Cresol, bandaids, phenol
Bitter	Fundamental taste sensation elicited by caffeine, quinine	Caffeine (0.08% in water)
Salty	Fundamental taste sensation elicited by salts	Sodium chloride (0.5% in water)
Sweet	Fundamental taste sensation elicited by sugars	Sucrose (5% in water)
Sour	Fundamental taste sensation elicited by acids	Citric acid (0.08% in water)
Umami	Chemical feeling factor elicited by certain peptides and nucleotides	MSG (1% in water)
Prickle	Chemical feeling factor of which the sensation of carbonation on the tongue is typical	Soda water

Source: MaryAnne Drake, Department of Food Science, North Carolina State University, Raleigh, NC. With permission.

III. Comparison of sensory analysis and analytical chemistry flavor analysis

To understand the strengths and weaknesses of sensory analysis and analytical chemistry flavor analysis (usually based on gas chromatography–mass spectrometry (GC-MS) analysis of extracts of volatiles and semivolatiles), it can be helpful to compare the basic principles of each approach. Although both sensory and analytical chemistry approaches are useful for resolving flavor issues, the basis of their methodologies is distinctly different.

With analytical flavor methods, flavor chemists attempt to imply how a food will taste, based on the most potent character-impact compounds (see Chapter 9) present in the sample, often relying on GC-MS and olfactometry results.

With QDA, panelists taste the food and try to assess the level of various taste attributes (e.g., salty, sour, vanilla, cheesy, etc.). Panelist results for samples are frequently compared with spider graphs to determine, for example, how closely a prototype sample matches a control sample, how competitor A sample matches competitor B sample, how a pass (acceptable) sample compares to a fail (rejected) sample, and so on. This type of sensory information can be used to guide attempts in making product formulation adjustments that yield products with more desirable and improved sensory attributes.

Sensory analysis is essentially a multivariate technique. A multivariate measurement is one in which multiple measurements are made on a sample of interest. The impact of all flavor chemicals present in a sample is interpreted *in toto*. The perceived flavor of a food sample, for example, involves the simultaneous tasting of all the flavor chemicals present. Sensory panelists cannot taste just one flavor chemical at a time; for some types of flavor research and for some types of samples, this can be a significant detriment.

To overcome the deficiencies of detecting all tastes simultaneously, sensory scientists use various techniques to make interpretation of sensory analysis results more univariate in nature. QDA is one approach. Sensory panelists may be asked to rank the degree of saltiness, sweetness, and vanilla taste of a baked product, based on a hedonic scale — for example, to assign a value of 0 for total lack of the flavor attribute and 10 for an intense impression of the flavor attribute. Quantifying chemicals responsible for specific flavor attributes in this way can be subjective, prone to error, and susceptible to statistical manipulation [16]. Disadvantages of sensory analysis include costs involved with training panelists, the inconvenience of assembling enough trained panelists to make sensory evaluations statistically significant, and the time required to perform sensory analysis.

If accurate assessment of the level of a particular flavor chemical is a weakness of sensory analysis when compared to chemical analytical testing, the ability to interpret the overall flavor impact of the myriad chemicals comprising a food is its strength.

Chemical analysis, in contrast, is essentially a univariate technique. Whereas hundreds of chemical peaks may appear in the chromatogram of a flavor extract of a food sample, interpretation of how the chemicals actually impact flavor is usually conducted by considering one chemical at a time. Inspecting a complicated chromatogram of a food extract does not usually reveal how the combination of flavor chemicals interacts on human taste buds in creating flavor sensation or what the ultimate taste of the product will be.

Whereas sensory scientists have developed techniques that attempt to make interpretation of their multivariate sensory analysis more univariate, analytical flavor chemists have developed ways to make interpretation of their univariate approach more multivariate. Although chemical data may appear to be multivariate, most data analysis software treats the data as a succession of noncorrelated, univariate measurements. To make interpretations of analytical data more multivariate in nature, analytical flavor chemists have learned to apply multivariate statistical analysis to their data. Today's GC-MS instruments can analyze dozens of samples, each with hundreds of chemical components. Although the software bundled with most instruments is not designed to extract meaningful information efficiently from such large data sets, new multivariate software packages now offer this capability. One example is the software developed by Gerstel GmbH & Co. KG (Mülheim an der Ruhr, Germany); the company has developed macros that can export Agilent's ChemStation data into the popular Piroutte® chemometrics software (Infometrix Inc., Woodinville, WA). This approach makes it possible to quickly and easily transfer extensive GC data files into chemometrics software for further data exploration and analysis. Compared to univariate regression, multivariate methods offer improved precision, more sophisticated outlier detection, and, in the case of factor-based algorithms, the possibility of compensating for interferences. These topics have also been previously described in the literature [17,18].

There are some analytical techniques that are based on multivariate measurements and multivariate interpretation of instrumental output. One example is the so-called electronic nose (e-nose) instrument that employs sensor arrays to obtain multiple responses on vapor samples. Although conducting polymer sensors and metal oxide sensors have been developed for measuring flavor volatiles in food systems, they have had limited success to date because of problems with sensor drift, the need for constant recalibration, and other factors. Perhaps the most successful and reliable e-nose instruments to date are those based on mass spectrometry detection [19].

IV.   Which is better?

Which is "better," analytical chemistry flavor analysis or sensory testing? The answer depends on the type of information sought. The key to resolving flavor problems is to match the appropriate tool (difference sensory testing, QDA, GC-MS analysis, olfactometry, etc.) to the specific problem being studied. If a food technologist has attempted to duplicate a competitor product and has created three prototypes to compare to the competitor, then sensory analysis is likely the preferred approach, with chemical analysis supplying a secondary supporting role. If a sample has developed an unfamiliar off-flavor immediately after production or during shelf life testing, then chemical analysis will likely be preferred over sensory approaches.

If the goal is to duplicate a flavor of a competitor product, a flavorist can often accomplish this without knowing the actual flavor chemicals involved. The flavorist can simply come up with the right combination of flavorants that mimics the flavor of the target product.

Often, chemical analysis can help sensory scientists get to the solution of the problem more quickly. Although many flavor problems can be solved without chemical analysis, there are times when an analytical chemistry approach is superior.

A.   Examples in which chemical analysis was better

1.   Sour cream off-flavor

A dairy technologist attempted to make a test batch of a new sour cream product in the pilot plant. His first attempt resulted in a product with a potent, unusual off-flavor, as did his second and third attempts. When samples were subjected to sensory analysis with panelists trained in tasting off-flavors in dairy products, the consensus opinion was that there was probably a problem with the culture bacteria used to make the product. However, when additional pilot-plant samples were prepared with two different types of cultures, the identical off-flavor problem developed. Finally, control and problem samples were analyzed by the flavor chemist. Results appear in Figure 1.1. Based on chemical analysis, oxidation of unsaturated fatty acids in butterfat was apparent as the cause of the problem. The chemist recommended inspection of the processing equipment. A new brass valve had recently been installed in a processing line. The valve was replaced with one made of stainless steel, and the problem disappeared.

This is an interesting example of how useful chemical analysis can be in resolving off-flavor issues. Sometimes sensory analysis totally fails in resolving a flavor problem. Even though they were well-trained in tasting dairy products (including identification of oxidation off-flavors), dairy technologists could not recognize oxidation off-flavors in these samples. Oxidation off-flavors commingled with flavors contributed by the metabolites from the culture bacteria created a combined unrecognizable flavor. This is an example where the univariate character of chemical analysis can be more advantageous than the multivariate character inherent with sensory analysis.

2.   Ice cream with burnt-feathers off-flavor

An ice cream manufacturer received several complaints that its vanilla ice cream from one warehouse had a severe off-flavor that was described as putrid, cabbage-like, or similar to burnt feathers. Samples from several different production runs developed the flavor defect, which was concentrated on the surface of the ice cream nearest the lid. The ice cream was packaged in a round 1.5-gal paperboard carton fitted with a clear plastic lid. As shown in Figure 1.2, analysis of complaint samples by purge-and-trap GC-MS

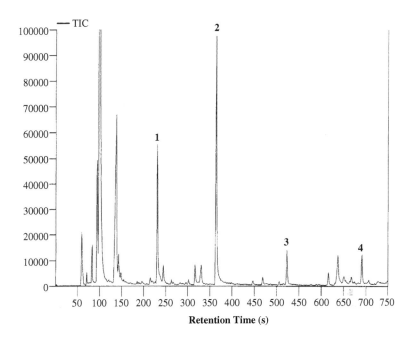

Figure 1.1 Sour cream with oxidation off-flavor. Sample analyzed by GC-MS using solid-phase microextraction (SPME). 1 = pentanal; 2 = hexanal; 3 = heptanal; 4 = octanal.

revealed that, compared to control samples, the complaint samples had significant concentrations of dimethyl disulfide and hexanal. It is well known that both of these compounds form in dairy products as a result of light-induced reactions: dimethyl disulfide involving methionine and hexanal from linoleic acid [20].

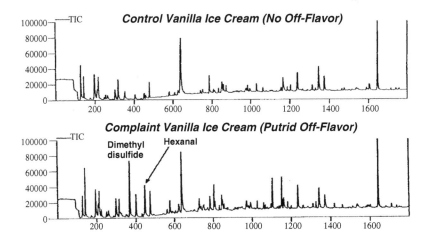

Figure 1.2 Burnt-feather off-flavor in ice cream.

Records showed that the flavor defect first became noticeable when samples were stored in a new warehouse. Careful inspection of the freezer warehouse storage conditions revealed that samples were stored in close proximity to high-intensity light. Lighting adjustments were made, and the problem was easily and quickly remedied.

This is an interesting case because light-induced off-flavors were generated in a food product while in a frozen state.

3. Meat analog with beany off-flavor

Another example shows how chemical analysis may be preferred over sensory analysis in off-flavor resolution studies. A food technologist had developed a meat analog product in the form of extruded bits. Ingredients included several artificial meat flavors, soy, and other vegetable-based components. The product had excellent textural and color qualities but had a strong "beany" flavor note that was unacceptable to the ingredient supplier's customer.

Because sensory analysis of the ingredients did not reveal the source of the contamination, the food technologist believed the off-flavor was created in thermal reactions during extrusion processing. Samples were sent to the analytical lab for GC-MS analysis. Results, shown in Figure 1.3, revealed a significant peak for 2-pentylfuran, an oxidation product of linoleic acid. 2-Pentylfuran, long associated with beany notes in soy products, was highly

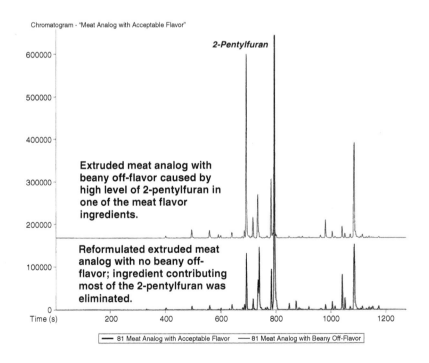

Figure 1.3 2-Pentylfuran in meat analog.

suspected as the cause of the off-flavor note. 2-Pentylfuran, an oxidation product of linoleic acid, could, indeed, be created during high-heat extrusion processing.

Using solid-phase microextraction (SPME) GC-MS techniques, chemists decided to screen all ingredients for the presence of 2-pentylfuran. One of the meat flavor ingredients contained a large peak for 2-pentylfuran. The ingredient was eliminated from product formulations, the beany off-note disappeared from the finished product, and the flavor of the new reformulation was enthusiastically accepted by the customer.

In this case, the multivariate nature of sensory analysis was a pitfall. Sensory panelists could not detect the distinct flavor of 2-pentylfuran in the complicated concoction of flavor chemicals in the artificial meat flavor. The univariate nature of chemical analysis was able to accurately measure the chemical causing the problem.

In these three examples, chemical analysis was able to resolve flavor problems that were not solved by sensory analysis. This is not to say, however, that chemical analysis is always superior to sensory analysis in resolving off-flavor issues. Indeed, there are times when sensory analysis can quickly resolve flavor issues that chemical analysis cannot readily explain.

B.   Limitations of chemical analysis

There are several potential limitations with chemical analysis:

1.   The character-impact chemicals may not be extracted by the sample preparation technique used prior to GC-MS analysis. All sample preparation techniques have biases. With purge-and-trap methods, for example, the odor active chemicals may be too volatile or not volatile enough to be trapped on the trapping medium used; if an SPME method is used, an inappropriate fiber may have been selected for the analytes of interest.

2.   The character-impact chemicals may be thermally labile and decompose in hot GC injectors. This problem has been reported for some types of volatile sulfur compounds. Flavor chemists sometimes use cool GC injectors to minimize this problem.

3.   The character-impact chemicals may be nonvolatile chemicals and should have been analyzed by HPLC, ion chromatography, or an analytical technique other than GC-MS. Examples of these types of chemicals include sugars, salts, lactic acid and other organic acids, amino acids, and peptides.

4.   The character-impact chemicals may be extracted with 100% recovery but are present at levels too low to be detected by GC detection. Some chemicals have extremely low odor or taste thresholds and profoundly impact the flavor of the food product but are present at concentrations below the detection level of the GC detector.

5.   Unexpected synergy problems sometimes arise with flavor volatiles. It is not always possible to predict the flavor impact of individual

components, even though they can be accurately measured by GC-MS. Sometimes when chemicals are smelled or tasted together, they are perceived differently than when they are smelled or tasted individually. Even with sophisticated, well-designed olfactometry studies, it can be difficult or impossible to predict which chemicals are responsible for the characteristic flavor of a food product.

C. Examples of the problem-solving capabilities of sensory analysis

Following are examples of how sensory analysis has been applied to solve significant flavor problems with dairy products.

1. Optimization of cheddar cheese taste in model cheese systems

Cheddar cheese, the most popular natural cheese in the U.S., has a very complex flavor system. Although much information has accumulated during the past century, dairy flavor chemists still lack a full understanding of cheddar cheese flavor and have not been able to replicate it in model systems. The nonvolatile sensory attributes of cheddar cheese are important for providing the character of cheddar cheese. Whereas much work has been published on volatile components of cheddar, far less is known on how nonvolatiles impact cheddar flavor. B. Yang and Z. Vickers used sensory analysis to better understand the importance of nonvolatile compounds to cheddar flavor [21].

A descriptive panel was trained to evaluate real and model cheese for a variety of taste attributes and for cheddarlike taste. Sodium chloride, lactic acid, citric acid, and monosodium glutamate were added to the model systems using mixture designs and response surface methodology to determine optimum levels of these components. The three model systems investigated were: (1) a dairy model system (containing milk isolate, anhydrous milk fat, water, annato color, and Chymosin), (2) a nondairy model system (containing gelatin, gum acacia, modified starch, sunflower oil, water, and annato color), and (3) a mozzarella base. Even though the mozzarella base did have taste, it was used because the other two model systems were too unlike cheddar cheese (or any cheese) in texture.

Less sodium chloride and acids were required to simulate the taste of mild cheddar compared with aged cheddar. None of the model cheese systems closely mimicked the texture of real cheddar. The researchers were able to match approximately, but not exactly, the taste of aged cheddar using a mozzarella base. Panelists generally rated the optimized taste in the dairy model system as more cheddarlike than the optimized tastes in the nondairy model.

Two methods were used to measure how close a sample was to the cheddar concept. One was by measuring the similarity of the sample to either mild cheddar cheese taste or aged cheddar cheese taste on an unstructured scale. The left end of the line was marked with "not at all like cheddar taste" and the right end was marked with "exactly like mild cheddar taste" or "exactly like aged cheddar taste." The other method was by concept matching using an R-index methodology. For the mild group, the panel evaluated whether the samples were MC (mild cheddar taste and sure), MC? (mild cheddar taste but not sure), N? (no cheddar taste but not sure) or N (no cheddar taste and sure). For the aged group, MC and MC? were changed to AC (aged cheddar taste and sure) and AC? (aged cheddar taste but not sure).

A model system for studying cheddar taste must be free of tastes and also have a texture and composition similar to that of a real cheddar cheese. The characteristic flavor of a food depends not only on the flavor compounds present and their levels but also the rate and extent to which they are released in real time, which in turn are affected by the amounts of proteins, fat, and other matrix components of the sample.

By using a trained descriptive analysis panel, Yang and Vickers were able to evaluate the flavor impact of several nonvolatile cheddar cheese components (i.e., salt, lactic acid, citric acid, and monosodium glutamate) in model systems that attempted to mimic real cheddar cheese. They achieved the most cheddarlike taste with the mozzarella cheese base, and panelists found the optimal concentration of salts and acids in the model to be nearly indistinguishable from real cheddar cheese.

In all three model systems and in water, more acids, more NaCl, and less MSG were needed to reconstruct aged cheddar taste than mild cheddar taste. This agrees with the observation that during aging, cheddar cheese becomes more intense in saltiness, sourness, and overall flavor intensity.

2. QDA and PCA for sensory characterization of ultrapasteurized milk

Extending the shelf life of fluid milk products will contribute to the competitiveness of the dairy industry in the beverage market. Ultrahigh temperature (UHT) processing and ultrapasteurization (UP) are two currently used approaches for extending dairy product shelf lives beyond those obtained by conventional pasteurization. One problem is that these products, which involve higher levels of heat treatment compared to conventional high-temperature–short-time (HTST) pasteurization, have been criticized for cooked off-flavors. As product flavor quality drives consumer acceptance and demand, the ability to measure sensory attributes that are characteristic of high-quality products is necessary for the development and production of products that meet consumer expectations.

K.W. Chapman et al. [22] used PCA of QDA results to identify and measure UP fluid milk product attributes that are important to consumers. The researchers studied nine UP milk products of various fat levels, including two lactose-reduced products, from two dairy plants. PCA identified four significant principal components that accounted for 94.4% of the variance in the sensory attribute data for UP milk samples. PCA scores indicated that the location of each UP milk along each of four scales primarily corresponded to cooked, drying or lingering, sweet, and bitter attributes. Overall product quality was modeled as a function of the principal components using multiple least squares regression (R2 = 0.810). These findings demonstrate

Table 1.2 Descriptors Used for Sensory Characterization of Ultrapasteurized Milk

Aroma	Flavor	Texture	Aftertaste
Cooked	Cooked	Viscosity	Drying
Caramelized	Sweet	Drying	Metallic
Grainy/malty	Caramelized	Chalky	Bitter
Other	Bitter	Lingering	Other
	Metallic		
	Other		

Source: From K.W. Chapman, H.T. Lawless, and K.J. Boor, Quantitative descriptive analysis and principal components analysis for sensory characterization of ultrapasteurized milk, *J. Dairy Sci.* 84: 12, 2001. With permission.

the utility of QDA for identifying and measuring UP fluid milk product attributes that are important to consumers.

The researchers were able to develop regression models that could be used to estimate the overall product quality rating based on measurement of its attributes. By plugging in QDA attribute scores for each sample, these regression equations could be used to calculate an overall quality rating for future samples tested. In general, perception of bitter flavor had the most dramatic effect on overall quality perception.

Table 1.2 lists the descriptors used for QDA, and Table 1.3 shows the Varimax rotated PC factor loadings for UP milk attributes. Figure 1.4 is a sensory profile for a reduced-fat UP milk sample stored at 6°C for 2 d (dark-grey area), 29 d (black area), and 61 d (light-grey area). Individual attributes are positioned like the spokes of a wheel around a center (zero or not detected) point, with the spokes representing attribute-intensity scales and higher (more intense) values radiating outward.

Table 1.3 Varimax Rotated Principal Component Factor Loadings for Ultrapasteurized Milk Attributes

Attributes	PC1	PC2	PC3	PC4
Cooked aroma	0.971[a]	0.013	0.034	−0.208
Caramel aroma	0.497	−0.539	−0.567[a]	−0.252
Grainy/malty aroma	0.964[a]	0.021	−0.231	0.032
Cooked flavor	0.702[a]	−0.547	0.091	−0.350
Sweet flavor	0.038	0.082	−0.969[a]	−0.146
Bitter flavor	−0.186	−0.003	0.191	0.946[a]
Dry texture	0.004	−0.942[a]	−0.101	−0.092
Lingering aftertaste	−0.003	−0.758[a]	0.389	0.413
Proportion of total variance	33.1%	25.7%	19.0%	16.6%

[a] Loading with an absolute value greater than 0.560; these attributes have the greatest sensory impact.

Source: From K.W. Chapman, H.T. Lawless, and K.J. Boor, Quantitative descriptive analysis and principal components analysis for sensory characterization of ultrapasteurized milk, *J. Dairy Sci.* 84: 12, 2001. With permission.

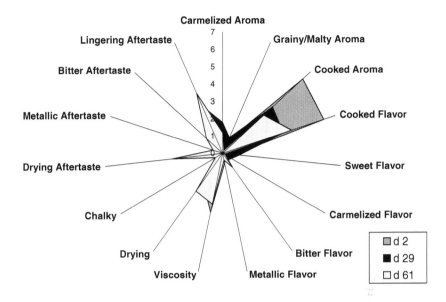

Figure 1.4 Sensory profile for a reduced-fat UP milk sample stored at 6°C for 2 d (dark-grey area), 29 d (black area), and 61 d (light-grey area). (From K.W. Chapman, H.T. Lawless, and K.J. Boor, Quantitative descriptive analysis and principal components analysis for sensory characterization of ultrapasteurized milk, *J. Dairy Sci.* 84: 12, 2001. With permission.)

3.   Acceptance of reduced-fat ultrapasteurized milk by consumers 6 to 11 years old

Milk products are commonly regarded as important dietary sources of protein, minerals, and vitamins. Despite these nutritional benefits, milk consumption by U.S. consumers typically bears an inverse relationship with consumer age, with the decline in per capita consumption beginning as early as ages 6 to 12, according to a 1999 Milk Industry Foundation study. In fact, milk consumption by 6- to 12-year-old children in 1997 was reported as 10% less than that consumed by 0- to 5-year-old children (Dairy Management Inc. Planning Research Group, 1999). In contrast, carbonated soft drink consumption was 176.8% greater among the 6 to 12 year olds than among the younger group. To increase the appeal of their UP and UHT milk offerings, dairy processors need to understand what flavor attributes are impacting flavor acceptance with this target audience and then devise ways to control these critical flavor attributes.

K.W. Chapman and K.J. Boor [23] studied the degree of liking of UP milk by 6 to 11 year olds. For comparative purposes, UP reduced-fat milks were evaluated along with conventionally pasteurized HTST reduced-fat milks and UHT reduced-fat milks. A seven-point facial hedonic scale with Peryam and Kroll verbal descriptors for affective testing with children was used with the 6-year-olds. For the older children, a seven-point hedonic scale with Peryam and Kroll verbal descriptors was used.

The mean degree of liking of UP milk was rated as slightly below "good" by children 6 to 11. HTST milk was liked slightly more than the UHT milk, which was liked slightly more than the UP milk. As UP milks are often distributed in fast food establishments, which are commonly frequented by children in this age group, attention should be directed toward making these products more appealing to children.

Figure 1.5 shows the distribution of ratings of milk, using a seven-point hedonic scale. (The black bar represents HTST milk, grey bar represents UP milk, and white bar represents UHT milk.) Although UP milks had a higher percentage of "good" scores than either HTST or UHT milk, the HTST and UHT milks had higher "really good" and "super good" percentages. How children felt about milk, in general, significantly affected how much they liked the test milks, with all types of milks being influenced equally.

This research is an excellent example of how sensory analysis can be used to understand the taste preferences of specific consumer target groups so products with appropriate sensory attributes can be developed for that group.

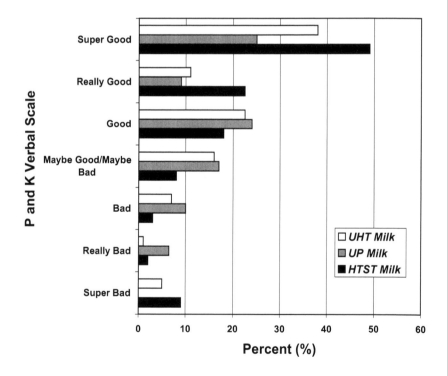

Figure 1.5 The distribution of ratings of milk, using a seven-point hedonic scale. (Black bar represents HTST milk, grey bar represents UP milk, and white bar represents UHT milk.) (From K.W. Chapman and K.J. Boor, Acceptance of 2% ultrapasteurized milk by consumers, 6 to 11 years old, *J. Dairy Sci.* 84: 951, 2001. With permission.)

4. Cheddar cheese and powdered-milk lexicons

M.A. Drake et al. [8], through the Southeast Dairy Foods Research Center, developed and validated a descriptive language for cheddar cheese flavor. Representative cheeses (240) were collected. Fifteen individuals from industry, academia, and government participated in roundtable discussions to generate descriptive flavor terms. A highly trained descriptive panel (n = 11) refined the terms and identified references. Identification of chemical references was conducted with the assistance of K. Cadwallader, University of Illinois. Cadwallader and coresearchers used GC-MS-olfactometry tests to identify many flavor compounds that were responsible for specific flavors and off-flavors in cheddar cheese.

Cheddar cheeses (n = 24) were presented to the panel for validation with the identified lexicon. The panel differentiated the 24 cheddar cheeses as determined by univariate and multivariate analysis of variance. Twenty-seven terms were identified to describe cheddar flavor. Seventeen descriptive terms were present in most cheddar cheeses. Drake's standard sensory language for cheddar cheese will facilitate training and communication among different research groups. The cheddar cheese lexicon can help cheesemakers and cheese users accurately and consistently characterize the flavor of their cheese products and improve quality issues by measuring and controlling the presence of those compounds that have been associated with flavor defects.

Following development of the cheddar cheese lexicon, Drake and coresearchers developed a similar language to help characterize another food industry staple — dairy protein powders, including whey proteins and nonfat dry milk [24].

Global production of nonfat dry milk tops 3.3 million tons (USDA 2000), and whey protein demand still outstrips production, which increases annually. A sensory lexicon describing the flavor of these ingredients helps dairy processors maximize the quality of these ingredients and allows food technologists to identify the exact attributes or flavor notes these ingredients contribute to formulations.

Drake was surprised by the number of descriptive terms that the panel uncovered for application to the milk protein powder lexicon. The panel discovered 21 flavor terms that could be applied to milk powders. Examples included cooked or milky flavor, cake mix or vanillin, sweet and sour, earth, cereal, and others. Each of these flavors was linked to a key aroma compound, many of which were identified by Cadwallader in GC-MS experiments. For example, lactones tend to lend a sweet, coconut-like flavor, whereas various free fatty acids can simulate a rancid or waxy flavor.

Many different factors contribute to flavor variability. The source of the powder, processing or packaging methods and materials, as well as storage time and conditions are just a few. The milk protein powder lexicon, linking responsible chemical factors and causal agents, provides common ground for processors and ingredient suppliers to discuss ingredient characteristics so quality improvement of these ingredients can be made.

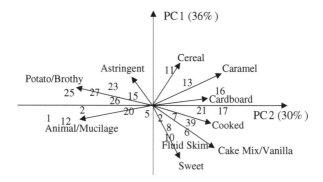

Figure 1.6 How QDA results based on Drake's protein powder lexicon can be analyzed by PCA. (From M.A. Drake, ADSA foundation scholar award: defining dairy flavors, *J. Dairy Sci.* 87: 777, 2004. With permission.)

Figure 1.6 shows how QDA results based on Drake's protein powder lexicon can be analyzed by PCA [25]. This two-dimensional PCA plot shows the attribute variability among 27 low-heat skim milk powders less than three months old. Rehydrated milk powders are represented by numbers (PC1 = principal component 1, PC2 = principal component 2).

V. Linking sensory and chemical analysis to better understand flavor development

As these few examples illustrate, sensory analysis and chemical analysis — each in its own right — are powerful problem-solving tools. However, coupling sensory analysis with chemical analytical data can provide even more insights than using either technique alone. Combining both sensory analysis and chemical analysis can be a potent tool for resolving many types of difficult flavor problems. The following examples describe how incorporating sensory input with analytical data can result in a powerful tool for problem solving.

A. Effect of antioxidant fortification on light-induced flavor of milk

Light-induced off-flavors are a major source of consumer complaints with processed milk. Oxidative reactions in milk are detrimental because they reduce the nutritional value of milk and also contribute to reduction in shelf life. M. van Aardt et al. [26] recently studied the effectiveness of added antioxidants against oxidation off-flavor development in light-exposed milk, using both sensory and chemical analyses.

Sensory testing for similarity showed no perceivable difference between control milk and milk with added 0.05% α-tocopherol (TOC) and 0.025% TOC plus 0.025% ascorbic acid (AA) but did demonstrate a perceivable difference with added 0.05% AA alone. Subsequently, sensory testing for difference showed a significant difference in oxidation off-flavor between light-exposed control milk and light-exposed milk with added TOC/AA, though milk fortified with only TOC was not different from the control. General remarks on scoresheets from panelists that correctly identified the "odd" samples indicated that reduced fat light-exposed milk treated with a combination of TOC/AA showed more fresh milk flavor character than light-exposed milk without added antioxidants. This implies that the significant difference that was observed between light-exposed control milk and milk treated with TOC/AA is because of a higher oxidized flavor in control milk. These results support previous studies that showed certain antioxidants such as TOC and AA demonstrate synergistic action against oxidation [27].

Researchers also examined samples by gas chromatography-olfactometry (GC-O), a technique involving extraction of flavor volatiles from the sample, injection of the extract into a heated GC injection port, separation of chemical components as they pass through a GC column, and finally sniffing of the individual chemical peaks as they elute from the column. It is interesting to note that light-exposed antioxidant-treated milk samples (TOC/AA and TOC) showed more odor active compounds than light-exposed control milk. Odor compounds such as 2-heptanone (nutty), 1-octanal (cooked, green, and grassy) and nonanal (sweet and plastic) were observed in antioxidant-treated samples. It is also interesting to note that, with light exposure, the addition of TOC seemed to increase the intensities of the aroma active compounds; this could indicate a pro-oxidant effect of the antioxidant. Although GC-O data suggested the presence of substantially more odorous flavor compounds in antioxidant-treated light-exposed milk, these compounds could be below human detection thresholds in the sample matrix, which might indicate why sensory results did not indicate increased light-oxidation flavor.

The thiobarbituric acid reactive substances (TBARS) test verified chemically the extent of oxidation in control and antioxidant-treated milk samples. Milk that was exposed to light for 10 h showed a significantly higher TBARS value (0.92 ± 0.09 mg/kg) than milk that was protected from light (0.59 ± 0.18 mg/kg) or milk that was treated with TOC/AA (0.26 ± 0.09 mg/kg).

Both sensory and chemical analyses showed that direct addition of low levels of a combination of 0.025% TOC (1.25% TOC per g fat) and 0.025% AA to reduced-fat milk protected milk flavor over 10 h of light exposure. Sensory analysis showed that TOC/AA was the only treatment that did not negatively impact fresh milk flavor and limited light-induced flavor in milk after 10 h of light exposure. On the other hand, addition of TOC to milk also did not negatively impact fresh milk flavor, but it did not limit oxidation flavor in milk that was exposed to light for 10 h.

B.   Characterization of nutty flavor in cheddar cheese

Cheese flavor is one of the most important criteria for determining consumer choice and acceptance. Aged cheddar flavor is characterized by sulfur, brothy, and nutty flavors. Research that elucidates the origin of the important nutty flavor notes in cheese is scarce.

Defining the sensory term "nutty" is a difficult task, as the aroma quality in all nuts is not exactly the same. With Drake's cheddar cheese lexicon, nutty flavor is defined as the "(nonspecific) nutlike aromatic associated with different nuts." Lightly toasted unsalted nuts, unsalted Wheat Thins® crackers, or roasted peanut oil extract were used as references for nutty flavor.

Identifying specific chemical compounds associated with particular flavors requires extensive and specific instrumental and sensory analysis. A three-step process was used by Y.K. Avsar et al. [28] to study the chemicals responsible for the nutty flavor of aged cheddar: (1) First, QDA was used to qualitatively and quantitatively identify all of the sensory-perceived flavors and tastes present in the cheese. (2) Instrumental (GC) techniques were then applied to identify volatile compounds that contribute to flavor. GC-O can assist in identification of compounds that are present in the sensory threshold range; it is often used as a way of further screening volatile compounds that play key roles in flavor. (3) Finally, model systems, similar to the actual cheese, were then constructed to evaluate the role of specific compounds on sensory-perceived flavor. This last step is sometimes referred to as recombination studies.

Using this process, researchers identified the key chemical components in cheese responsible for nutty flavor notes. Sensory analysis of cheese models revealed that three Strecker aldehydes — 2-methylpropanal, 2-methylbutanal and 3-methylbutanal — can contribute to nutty flavors in aged (> 9 months) cheddar cheeses. Quantitative data suggested that 2-methylpropanal may be most important, because it was more prevalent in nutty cheese and present at higher concentrations than the other Strecker aldehydes.

The formation of these aldehydes requires the presence of certain amino acids — valine for 2-methylpropanal, isoleucine for 2-methylbutanal, and leucine for 3-methylbutanal. In order to produce cheddar cheese with enhanced or accelerated nutty flavor, the researchers advised using one of the following three methods: (1) the use of starter bacteria capable of releasing these certain amino acids, (2) addition of certain amino acids into cheese milk or cheese slurry, and (3) accelerating the conversion rate of these amino acids into aroma compounds (Strecker aldehydes).

This study is an excellent example of how combining sensory and analytical studies can be used to formulate cheeses with specific flavor qualities that might be of interest in specific applications or to specific market segments.

References

1. L.G. Phillips, M.L. McGiff, D.M. Barbano, and H.T. Lawless, The influence of nonfat dry milk on the sensory properties, viscosity and color of lowfat milks, *J. Dairy Sci.* 78: 1258, 1995.
2. H.J. Quinones, D.M. Barbano, and L.G. Phillips, Influence of protein standardization by ultrafiltration on the viscosity, color and sensory properties of 2% and 3.3% milk, *J. Dairy Sci.* 81: 884, 1998.

3. R.L. Ohmes, R.T. Marshall, and H. Heymann, Sensory and physical properties of ice creams containing milk fat or fat replacers, *J. Dairy Sci.* 81: 1222, 1998.
4. A.M. Roland, L.G. Phillips, and K.J. Boor, Effects of fat replacers on the sensory properties, color, melting and hardness of ice cream, *J. Dairy Sci.* 82: 2094, 1999.
5. A.I. Ordonex, F.C. Ibanez, P. Torre, Y. Barcina, and F.J. Perez-Elortondo, Application of multivariate analysis to sensory characterization of ewes' milk cheese, *J. Sensory Stud.* 12: 45, 1998.
6. H.T. Lawless and H. Heymann, *Sensory Evaluation of Food: Principles and Practices*, Chapman and Hall, New York, 1998, p. 606.
7. M.A. Drake and G.V. Civille, Flavor lexicons, *Compr. Rev.Food Sci. Food Saf.* 2: 33–39, 2002.
8. M.A. Drake, S.C. McIngvale, K.R. Cadwallader, and G.V. Civille, Development of a descriptive sensory language for Cheddar cheese, *J. Food Sci.* 66: 1422, 2001.
9. D.D. Muir, E.A. Hunger, J.M. Banks, and D.S. Horne, Sensory properties of hard cheeses: identification of key attributes, *Int. Dairy J.* 5: 157, 1995.
10. J.R. Piggot and R.G. Mowat, Sensory aspects of maturation of Cheddar cheese by descriptive analysis, *J. Sensory Stud.* 6: 49, 1991.
11. A.K. Roberts and Z.M. Vickers, Cheddar cheese aging: changes in sensory attributes and consumer acceptance, *J. Food Sci.* 59: 328, 1994.
12. D.D. Muir, J.M. Banks, and E.A. Hunter, Sensory properties of Cheddar cheese: effect of starter type and adjunction, *Int. Dairy J.* 6: 407, 1996.
13. M.A. Drake, T.D. Boylston, K.D. Spence, and B.G. Swanson, Chemical and sensory quality of reduced fat cheese with a *Lactobacillus* adjunct, *Food Res. Int.* 29(3/4): 381, 1996.
14. M.A. Drake, T.D. Boylston, K.D. Spence, and B.G. Swanson, Improvement of sensory quality of reduced fat Cheddar cheese by a *Lactobacillus* adjunct, *Food Res. Int.* 30(1): 35, 1997.
15. J.M. Banks, E.A. Hunter, and D.D. Muir, Sensory properties of low fat Cheddar; effect of salt content and adjunct culture, *J. Soc. Dairy Technol.* 46: 119, 1993.
16. J. Wright, *Flavor Creation*, Allured Publishing Corporation, Carol Stream, IL, 2004, p. 195.
17. M. Chien and T. Peppard, Use of statistical methods to better understand gas chromatographic data obtained from complex flavor systems, *Flavor Measurement*, C.-T. Ho and C.H. Manley, Eds., Marcel Dekker, New York, 1993, p. 1.
18. K.R. Beebe, R.J. Pell, and M.B. Seaholtz, *Chemometrics: A Practical Guide*, John Wiley & Sons, New York, 1998.
19. R.T. Marsili, Combining mass spectrometry and multivariate analysis to make a reliable and versatile electronic nose, *Flavor, Fragrance and Odor Analysis*, R. Marsili, Ed., Marcel Dekker, New York, 2002, p. 349.
20. R.T. Marsili, Flavours and off-flavours in dairy foods, *Encyclopedia of Dairy Sciences*, Vol. 2, H. Roginski, F.W. Fuquay, and P.F. Fox, Eds., Academic Press, London, 2003, p. 1069.
21. B. Yang and Z. Vickers, Optimization of cheddar cheese taste in model cheese systems, *J. Food Sci.* 69: S229, 2004.
22. K.W. Chapman, H.T. Lawless, and K.J. Boor, Quantitative descriptive analysis and principal components analysis for sensory characterization of ultrapasteurized milk, *J. Dairy Sci.* 84: 12, 2001.

23. K.W. Chapman and K.J. Boor, Acceptance of 2% ultrapasteurized milk by consumers, 6 to 11 years old, *J. Dairy Sci.* 84: 951, 2001.
24. M.A. Drake, Y. Karagul-Yuceer, K.R. Cadwallader, G.V. Civille, and P.S. Tong, Determination of the sensory attributes of dried milk powders and dairy ingredients, *J. Sensory Stud.* 18: 199, 2003.
25. M.A. Drake, ADSA foundation scholar award: defining dairy flavors, *J. Dairy Sci.* 87: 777, 2004.
26. M. van Aardt, S.E. Duncan, J.E. Marcy, T.E. Long, S.F. O'Keefe, and S.R. Nielsen-Sims, Effect of antioxidant fortification on light-induced flavor of milk, *J. Dairy Sci.*, 88: 872–880, 2005.
27. D.L. Madhavi, S.S. Deshpande, and D.K. Salunkhe, Introduction, *Food Antioxidants: Technological, Toxicological and Health Perspectives*, D.L. Madhavi, S.S. Deshpande, and D.K. Salunkhe, Eds., Marcel Dekker, New York, 1996, p. 1.
28. Y.K. Avsar, Y. Daragul-Yceer, M.A. Drake, T.K. Singh, Y. Yoon, and K.R. Cadwallader, Characterization of nutty flavor in Cheddar cheese, *J. Dairy Sci.* 87: 1999, 2004.

chapter two

Relating sensory and instrumental analyses

M.A. Drake, R.E. Miracle, A.D. Caudle, and K.R. Cadwallader

Contents

I. Introduction

Understanding food flavor is crucial to effective and strategic research and marketing. Understanding flavor partly involves linking the sensory perception of flavor and its volatile chemical components. Establishing these relationships requires sensory, as well as instrumental analysis, and can be challenging. Two basic approaches for relating data from sensory and

23

instrumental analyses are discussed. Examples are presented with butter and skim milk powder.

Food flavor is a key parameter for consumer acceptance and marketing. When specific links are definitively established between volatile compounds and sensory perception, an enhanced understanding of flavor is achieved, and powerful information for linking flavor to production technology is obtained (Drake, 2004). There are two basic requirements for such studies: complete instrumental analysis and descriptive sensory analysis.

Both descriptive sensory analysis and instrumental volatile compound extraction and identification techniques are reviewed in detail elsewhere (McGorrin, 2001; Drake and Civille, 2002; Singh et al., 2003; Drake, 2006). Careful attention must be paid to both techniques in order to establish clear relationships. Unfortunately, descriptive sensory analysis is often overlooked in flavor chemistry. Flavor is a sensory perception. Flavor chemistry research (e.g., instrumental analysis) has no relevance to flavor without sensory analysis. Descriptive sensory analysis consists of a trained panel in which individuals function in unison, analogous to components of an instrument, to document and describe the sensory attributes of a product (Drake and Civille, 2003). As such, the sensory instrument or panel, must receive extensive training and calibration in order to produce powerful, sensitive, and meaningful results. Selection of the appropriate volatile compound extraction approach is crucial as no single approach will extract all volatile components with equal recovery; so a combination of techniques (e.g., headspace vs. solvent extraction) should be used if a complete picture of volatile compounds is desired. Instrumentation must be sensitive as many flavor-contributing compounds can be present at low concentrations (ppb or ppt). Sulfur and nitrogen-containing compounds may require specific detectors to obtain accurate and sensitive detection and quantitation. Gas chromatography-olfactometry (GC-O or GC-sniffing) is generally necessary in order to identify key flavor-contributing compounds (Singh et al., 2003; Parliament and McGorrin, 2000; Van Ruth, 2001; Grosch, 1993).

The most precise and labor-intensive approach to linking sensory and analytical data consists of three basic steps: (1) selection of products with the desired or target flavors using descriptive sensory analysis, (2) instrumental volatile analysis, and (3) confirmation of key aroma-active volatile compounds using quantitation, threshold analysis, and descriptive sensory analysis of model systems (Drake et al., 2006). However, other more general approaches may be used when large numbers of samples or volatile components are evaluated. These alternative approaches would include (1) sensory evaluation and instrumental analysis of products without further analyses or (2) sensory evaluation, instrumental analysis and statistical analysis. In these two approaches, gross relationships can be identified. Specific links cannot be established, but there are many cases where such studies are the necessary starting point in order to facilitate additional and more detailed studies. This chapter will address these approaches with specific product research examples.

II. Establishing precise links: Identification of an off-flavor in butter as an example

The precise three-step approach (Figure 2.1) to establishing specific linkages between sensory perception and volatile components has been applied to many foods from bread to strawberries and to specific flavors in cheese and whey proteins (Scheiberle and Hofman, 1997; Kirchhoff and Scheiberle, 2001; Rychlik and others, 1998; Avsar et al., 2004; Carunchia-Whetstine et al., 2005; Drake et al., 2006; Wright et al., 2006). This approach is labor-intensive but establishes clear linkages. It works optimally when a specific flavor is targeted (e.g., nutty flavor in Cheddar cheese) and when one or a few compounds are responsible for a particular flavor. An example of this approach using butter follows.

Recently, a commercially produced butter was recalled due to an off-flavor. Qualitative tests indicated that there were no deviations in butter composition and that the fluid-milk source also displayed the off-flavor. Our goal was to identify the compounds responsible for the off-flavor. The three-step approach was used (Figure 2.1).

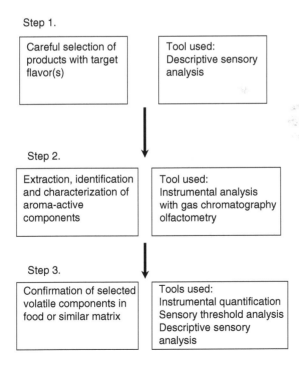

Step 1.

| Careful selection of products with target flavor(s) | Tool used: Descriptive sensory analysis |

Step 2.

| Extraction, identification and characterization of aroma-active components | Tool used: Instrumental analysis with gas chromatography olfactometry |

Step 3.

| Confirmation of selected volatile components in food or similar matrix | Tools used: Instrumental quantification Sensory threshold analysis Descriptive sensory analysis |

Figure 2.1 Three-step approach for establishing precise links between sensory and instrumental analyses.

A.   Materials and methods

Butters. Two blocks (22 kg) of commercial butter (> 2 months old, un-salted, fat content: 80.5 + 0.23%) were collected and profiled by a highly trained, descriptive, sensory analysis panel (n = 7). Each pan-elist had received more than 350 h of training using the Spectrum™ descriptive sensory analysis method (Meilgaard et al., 1999) (Sensory Spectrum, Chatham, NJ), and had extensive experience with the de-scriptive analysis of dairy products. Panelists profiled the butter us-ing an identified and defined sensory language (Table 2.1). Butters were presented in 3-digit coded sample cups with lids and tempered to 15°C prior to sensory analysis. Each butter was evaluated in du-plicate by each panelist using computerized data entry (Compusense 5 v 4.6, Compusense, Guelph, Canada). Data were analyzed by anal-ysis of variance with means separation (SAS Statistical Analysis Soft-ware, version 9.1, SAS Institute, Cary, NC).

Chemicals. Ethyl ether (anhydrous, 99.8%), sodium chloride (99%), sodium sulfate (99%), 2-methyl-3-heptanone and 2-methyl pentanoic acid (internal standards for neutral/basic fraction and acidic fractions, respectively) were obtained from Aldrich Chemical Company (St. Louis, MO). Aroma compounds that were positively identified were provided by Aldrich Chemical Co., Lancaster, (Windham, NH), and TCI America, (Portland, OR). Sodium bicarbonate (99.7%, w/w) and hydrochloric acid (36.5%, w/w) were obtained from Fisher Scientific (Pittsburgh, PA).

Direct solvent extraction. Butters were stored frozen (–80°C) and grated using a hand grater prior to extraction. Freshly grated butter (80 g) was weighed and divided among four Teflon bottles (capacity of 250 ml) with Tefzel™ closures in duplicate (160 g total for each butter) (Nalgene, Rochester, NY). Ethyl ether (80 ml) and 10 μl of internal standard (50 μl of 2-methyl-3-heptanone and 50 μl of 2-methyl pen-tanoic acid in 5 ml of methanol) were added to each bottle. The mixture was shaken for 30 min on a Roto mix (Thermolyne, type 50800; Dubuque, IA) at high speed. The bottles were then centrifuged

Table 2.1 Descriptive Sensory Language for Butter

Descriptor	Definition
Diacetyl/cultured	Aromatic associated with diacetyl
Milkfat/lactone	Aromatic characteristic of milk fat and coconut
Cooked/nutty	Aromatic associated with cooked milk and canned corn
Refrigerator/stale	Aroma characteristic of refrigerator with old food left in it, not one specific flavor but generally suggestive of a lack of freshness
Fatty/frier oil	Aromatic characteristic of stale fats and old frier oils
Salty taste	Basic taste associated with salts, such as sodium chloride

at 735 g for 10 min in order to separate the solvent phase from the mixture, which was subsequently collected into a glass jar. The procedure was repeated twice with 50 ml of ethyl ether. The solvent phases were combined and kept at −20°C overnight.

Solvent assisted flavor evaporation (SAFE). Volatile compounds from butter extracts were distilled using SAFE (Ace Glassware; Vineland, NJ). The assembly used was similar to that described by Engel et al. (1999). The SAFE apparatus was connected to a receiving tube and a waste tube. The glassware was then connected to a rough pump or diffusion pump as the vacuum source. The receiving tube and waste tube were held in separate Dewar flasks containing liquid nitrogen at all times. Distillation was carried out for 2 h under vacuum (ca. 10^{-4} Torr). Liquid solvent extract was loaded into the top of the SAFE apparatus, and released into the vacuum drop wise until all the extract had been placed under vacuum conditions. The SAFE apparatus was kept thermostated at 50°C with a circulating water bath. After distillation, the distillate was concentrated to 20 ml under a stream of nitrogen gas. Concentrated distillate was then washed twice with 3 ml sodium bicarbonate (0.5 M) and vigorously shaken. This step raises the pH of the extracts to ~11.0 and helps to separate the neutral/basic compounds from the acidic compounds. The distillate was then washed three times with 2 ml saturated sodium chloride solution in order to remove any residual water. After each wash step, the solution was shaken and the upper layer (ether) containing the neutral/basic fraction was collected using a pipette. The upper (neutral/basic) layers were then pooled, and the extract was dried over anhydrous sodium sulfate and concentrated to 0.5 ml under a stream of nitrogen gas. Acidic volatiles were recovered by acidifying the bottom layer (aqueous phase) with hydrochloric acid (18%) to 2 to 2.5 pH and extracting the sample three times with 15 ml ethyl ether. The acidified extract was dried over anhydrous sodium sulfate before concentration to 0.5 ml under a nitrogen gas stream.

Gas chromatography-olfactometry (GC-O) and AEDA. The semiquantitative GC-O technique, aroma-extract dilution analysis (AEDA) was used to characterize aroma-active compounds contributing to flavor in butters (Grosch, 1993; Van Ruth, 2001). AEDA was performed using a HP5890 series II gas chromatograph (Hewlett Packard Co., Palo Alto, CA) equipped with a flame ionization detector (FID), and a sniffing port and splitless injector were used. Both the neutral/basic and acidic fractions were analyzed from every extraction, and 2 µl was injected into a polar capillary column (DB-WAX 30 m length × 0.25 mm ID × 0.25 µm film thickness df; J & W Scientific, Folson, CA) and a nonpolar column (DB-5MS 30 m length × 0.25 mm ID × 0.25 µm df; J & W Scientific, Folson, CA). Column effluent was split 1:1 between the FID and the sniffing port using deactivated fused silica capillaries (1 m length × 0.25 mm ID). The GC oven temperature was programmed from 40 to 200°C at a rate of 10°C/min, with an initial hold for 3 min and a final hold of 20 min. The FID and sniffing port were maintained at a temperature of 250°C. The sniffing port was supplied with humidified air at 30 ml/min. The extracts were diluted stepwise with diethyl ether at a ratio of 1/3 (v/v). Two experienced sniffers, each with greater than 50 h training on GC-O, were used for AEDA. The dilution procedure was followed until sniffers detected no odorants. The highest dilution was reported as the $\log_3$ flavor dilution (FD) factor (Grosch, 1993).

Gas chromatography-mass spectrometry (GC-MS). For GC-MS analysis of the solvent extracts, an Agilent 6890 with 5973N mass-selective detector was used. Separations were performed on a fused silica capillary column (DB-5MS 30 m length × 0.25 mm ID × 0.25 µm df; J & W Scientific, Folson, CA). Helium gas was used as a carrier at a constant flow of 1 ml/min. Oven temperature was programmed from 35 to 200°C at a rate of 5°C/min, with initial and final hold times of 3 and 24 min, respectively. MSD conditions were as follows: MS transfer line heater, 250°C; MS Quad, 150°C; MS Source, 250°C; ionization energy, 70 eV; mass range, m/z 50 to 300; scan rate, 5 scan/sec. Each extract (1 µl) was injected in the splitless mode of a split or splitless injector at 250°C with a purge time of 1 min. Duplicate analyses were performed on each sample. Based on MS results, relative concentrations of the compounds were calculated. The area ratio (area of internal standard/area of compound) was multiplied by the concentration of the internal standard to determine the relative abundance of the compound. For positive identifications, retention indices (RI), mass spectra, and odor properties of unknowns were compared with those of authentic standard compounds analyzed under identical conditions. Tentative identifications were based on comparing mass spectra of unknown compounds with those in the National Institute of Standards and Technology (NIST05), mass spectral database or on matching the RI values and odor properties of unknowns against those of authentic standards. For the calculation of retention indices, an n-alkane series was used (Van den Dool and Kratz, 1963).

Threshold determination of BMS. Best-estimate thresholds of benzyl methyl sulfide (BMS) were determined using the ASTM ascending forced-choice method of limits procedure E679–79 (ASTM, 1992). The threshold of BMS was determined orthonasally in vegetable oil. Stock solutions of BMS were prepared in methanol. Aliquots of these stock solutions were placed into vegetable oil. These solutions were serially diluted (factor of 3) and 15 ml of each was poured into clean, labeled 56-ml plastic cups. Panelists (n = 25) were given these concentrations in a series, with two vegetable oil blanks containing methanol. Seven ascending series were tested. Series were presented in ascending concentration (4.3, 12.9, 38.8, 116.3, 348.8, 1046.5, and 3139.4 ppb). Each series was presented in a randomized order and evaluated by panelists.

Subjects were briefly instructed prior to testing. They were told to open the cups and briefly sniff the headspace of each cup in the series. Subjects rested 1 min between each set of three and were also instructed to sniff their sleeve to assist cleaning their nasal passageways between cups. The individual best-estimate threshold was calculated as the geometric mean of the last concentration, with an incorrect response, and the first concentration with a correct response. The group best-estimate threshold (BET) was calculated as the geometric mean of the individual best-estimate thresholds.

Sensory evaluation of butter models. Butter models were prepared using a commercial unsalted butter (purchased from a local grocery store) that did not display off-flavors (as determined by descriptive sensory analysis). BMS stock solution was prepared in 95% ethanol (flavor and aroma evaluation) at concentrations of 2, 20, 150, 300, and 600 ppb. These concentrations were chosen to bracket the possible concentration range identified in butter exhibiting the off-flavor. The butter was melted at 45°C and the stock solution introduced by a clean, disposable micropipet. After addition of the stock solution, butter models were swirled for 3 min and portioned (~15 g) into sample cups and then equilibrated for 24 h at 5°C. Butter models were evaluated for aroma or flavor by sensory analysis, using the same procedure applied for descriptive analysis of butter.

Quantification of BMS in butter using solid-phase microextraction (SPME). Butter (10 g) was transferred into precleaned 40 ml amber vials with screw caps and Teflon silica septa (Supelco, Bellefonte, PA). A magnetic octagonal stirring bar (8 mm OD × 13 mm length; Fisher, Pittsburgh, PA) was added. Each sample was allowed to equilibrate in the vial at 80°C for 15 min using a heating or stirring module (Reacti-Therm™, Pierce Biotechnology, Inc., Rockford, IL). Volatile odor compounds were extracted by exposing a 2 cm–50/30 μm divinylbenzene/carboxen/polydimethylsiloxane (DVB/CAR/PDMS) StableFlex SPME fiber (Supelco, Bellefonte, PA) to the headspace of each sample at 80°C for 60 min at a depth of 4 cm. The fiber was desorbed at 270°C for 10 min in the injection port fitted with a SPME inlet liner (Supelco, Bellefonte, PA) at a depth of 3 cm on a Varian CP-3380 GC/Saturn 2000 ion-trap mass-selective detector (Varian Inc.; Palo Alto, CA). Separations were performed on a fused silica capillary column (Rtx–5 30 m length × 0.25 mm ID × 0.25 μm df (Restek, Bellefonte, PA). Helium gas was used as a carrier at a constant flow of 1 ml/min. The oven temperature was programmed from 40 to 200°C at a rate of 5°C/min with initial and final hold times of 5 and 13 min, respectively. MSD conditions were as follows: transfer line 120°C; manifold 80°C; ion trap 150°C; ionization via automatic electronic ionization; mass range, m/z 35 to 350; EM voltage 2135V; scan/sec = 1. Standard addition was used to build a standard curve to determine the concentration of BMS in the off-flavored butter. A five-point curve was produced in duplicate with final additions of 0, 125.6, 251.2, 502.4, and 1204.8 ppb of BMS to samples of the off-flavored butter. A regression line through these points yielded the starting concentration of BMS in the sample as the x-axis intercept.

B. Results

The off-flavor in the butter was described as burnt/rubbery/sulfurous by trained sensory panelists (Figure 2.2). Other flavors and tastes in the off-flavored butter were not different from the control or reference butter. A clear sensory profile using sensory terms that can be readily understood and reproduced (Table 2.1), provides a firm foundation for identification of the off-flavor. Identification of the chemical sources of this off-flavor, or one of them, provides a method to not only identify the source of the off-flavor but also to readily train individuals to recognize and scale the off-flavor (Drake and Civille, 2003).

   A comparison of the GC-O and AEDA results revealed differences between the control and off-flavored butter (Table 2.2). The butters were characterized by an array of aroma-active compounds ranging from free fatty acids to lactones. Many of these compounds have been previously found in butter (Schieberle et al., 1993; Peterson and Reineccius, 2003a, 2003b). Most notably, a compound with a "burnt" or "rubbery" aroma was detected at moderate intensities in the neutral/basic fraction of the off-flavored butter. The compound had a $\log_3$ FD of 3 in the off-flavored butter and was not detected in the control butter. Mass spectral analysis identified the compound as BMS, and its identity was confirmed by mass spectrophotometry of the reference compound. These results (compound present at moderate intensity in the off-flavored product and absent in the control product, aroma similar to off-flavor) provide compelling evidence that BMS is the source of the off-flavor. However, GC-O data is not quantitative, and the aroma of an isolated compound is not always indicative of its role in flavor (Drake and Civille, 2003;

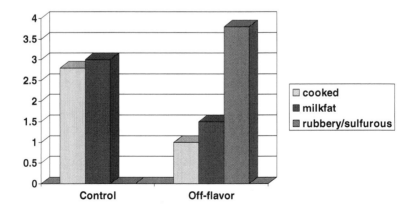

Figure 2.2 Sensory profiles of butters.

Table 2.2 Aroma Active Compounds of Butter

Number	Compound	Fraction	Butter control	Butter off	Odor[a]	RI[b] DB-5	DB-wax
			Mean Intensity (FD Factor)[c]				
1	Unknown	NB	1.50 (1)		Garbage sweet	698	999
2	(Z)-2-Penten-1-ol	NB	1.00 (1)	3.50 (<1)	Spicy sweet solvent	766	1120
3	Isobutyl acetate	NB	1.00 (1)		Sweet	776	1014
4	Dimethyl disulfide	NB		1.50 (<1)	Onion weedy	770	1063
5	Unknown	NB	1.25 (3)	2.00 (<1)	Sweet solvent plastic bottle	789	1111
6	Butyl acetate	NB	1.00 (3)		Fruity	831	
7	Unknown	NB	2.00 (1)	3.00 (1)	Potato skunk	835	1170
8	Isopropyl butanoate	NB	2.25 (4)	1.00 (<1)	Fruity bubblegum sweet solvent	844	1195
9	Ethyl 3-methylbutanoate	NB	2.25 (4)	3.00 (4)	Spicy solvent sweet fruity	864	
10	Methylfuranthiol	NB	2.00 (1)	3.00 (<1)	Rotten oil sulfur	868	
11	Heptanal	NB	2.00 (1)	2.00 (<1)	Fatty	907	
12	Methional	NB	1.33 (1)	2.50 (1)	Potato	923	
13	α-pinene	NB	1.50 (3)	3.00 (<1)	Urine mint	935	1029
14	Dimethyl trisulfide	NB	2.00 (<1)	1.88 (1)	Garlic cabbage sulfur	978	1383
15	(Z)-1,5-octadien-3-One	NB	2.00 (3)	3.00 (2)	Metallic earthy mushroom	983	1271
16	Unknown	NB	2.00 (2)		Citrus sweet fruity	1009	1352
17	D-limonene	NB	1.00 (4)		Sweet musty citrus	1037	
18	2-Nonanone	NB	3.50 (3)	2.63 (3)	Dusty musty geranium earthy	1094	
19	Nonanal	NB	4.00 (3)	2.50 (1)	Doughy fatty earthy	1107	1389
20	Unknown	NB	2.00 (4)		Citrus chemical	1113	
21	(E,Z)-2,6-Nonadienol	NB	2.75 (3)		Sweet cucumber	1171	
22	Benzyl methyl sulfide	NB		3.00 (3)	Sulfur rubber	1189	1590
23	Unknown	NB	2.00 (<1)		Mint cilantro	1198	
24	(E,E)-2,4-decadienal	NB	3.00 (3)		Fatty	1316	

(continued)

Sensory-directed flavor analysis

Table 2.2 (Continued) Aroma Active Compounds of Butter

Number	Compound	Fraction	Butter control	Butter off	Odor[a]	RI[b] DB-5	DB-wax
			Mean Intensity (FD Factor)[c]				
25	R-δ-Decalactone	NB	3.00 (2)	3.00 (<1)	Coconut butter fatty sweet cream	1438	2209
26	Unknown	NB	1.00 (4)		Dusty solvent cleanser	1502	
27	δ-Decalactone	NB	2.25 (2)	2.00 (<1)	Sweet waxy pecan coconut	1530	2213
28	Unknown	NB	4.00 (2)	3.50 (2)	Coconut	1567	
29	Unknown	NB	3.00 (3)		Coconut	1592	
30	δ-Undecalactone	NB	1.00 (<1)	2.00 (4)	Waxy peach sweaty	1617	
31	(Z)-Whiskey lactone	NB	3.00 (4)	1.00 (<1)	Fresh cream burnt sulfur	1645	1988
32	γ-Dodecalactone	NB	1.50 (3)	2.00 (1)	Coconut peach	1692	
33	δ-Dodecalactone	NB	4.50 (<1)	2.00 (<1)	Butter fatty sweet cream peach	1767	
34	Unknown	AC	2.00 (<1)	2.00 (<1)	Acidic sweat banana coconut	919	1095
35	Heptanone	AC	1.50 (<1)	3.50 (<1)	Spicy sweaty skunky acidic	903	1189
36	Trimethylthiazole	AC	2.50 (1)	1.50 (<1)	Potato fatty musty dusty		1379
37	Propanoic acid	AC	2.50 (1)		Sweaty swiss	668	1489
38	Isobutyric acid	AC	3.00 (3)	2.50 (<1)	Butyric sweaty cheesy	1227	1541
39	Butyric acid	AC	3.00 (4)	3.50 (1)	Sweaty cheesy animal fecal	838	1593
40	Isovaleric acid	AC	3.50 (4)	4.00 (3)	Sweaty cheesy fruity apricots	872	1696
41	Hexanoic acid	AC	2.00 (2)	1.00 (<1)	Waxy sweaty cheesy	1004	1814
42	Pantolactone	AC	2.00 (4)		Sweet burnt sugar		2013
43	Octanoic acid	AC	1.00 (<1)		Cheesy	1272	2082
44	γ-Decalactone	AC	2.00 (1)	2.50 (1)	Brothy oxidized lotion		2104

[a] Odor description at the GC-sniffing port.

[b] Retention indices were calculated from GCO data.

[c] Mean intensities (n = 2) and flavor dilution factors (log 3) were determined on a DB-5 column for NB compounds, and on a DB-Wax column for Ac compounds.

Drake et al., 2006). Thus, to confirm BMS as a source of the off-flavor, additional sensory and instrumental work was conducted.

Orthonasal threshold analysis revealed that the BET for BMS was 177 ± 98 ppb. The BET was calculated in oil rather than water since oil best represents the matrix most similar to butter. Simultaneously, the concentration of BMS in the off-flavored butter was calculated at 354 ± 60 ppb or 17% RSD (relative standard error). BMS was not detected in the control butter (by GC-O or GC-MS). The concentration of BMS in the off-flavored butter is above the sensory threshold, providing further evidence that BMS is the source of the off-flavor. The final step involves sensory analysis of model systems to show that the addition of the compound in the concentration range found in the off-flavored butter results in a flavor or aroma similar to the off-flavor in the butter (Figure 2.3). Addition of BMS to melted butter resulted in a burnt rubbery aroma and flavor similar to the aroma and flavor of the original off-flavored butter. Taken as a whole, these results are compelling evidence that BMS is the source of the off-flavor. Future steps would include identification of the source of BMS in butter (e.g., fluid milk source or storage-related). It is important to note that identification of flavor sources is not necessarily as straightforward as the example provided. A flavor may not be caused by a single compound, or the aroma of individual compound(s) by GC-O and may not be indicative of the flavors they contribute in the food product (Avsar et al., 2004; Drake et al., 2006). In such cases, more extensive analyses may be required. These analyses may involve calculation of odor activity values (OAVs) or more extensive sensory model addition and subtraction studies.

The odor activity value is the ratio of concentration (in the food of interest) to the sensory odor threshold for a particular compound (Nursten and Reineccius, 1996). OAVs may provide further information on which

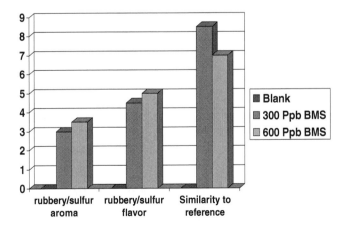

Figure 2.3 Intensity of rubbery or sulfur aroma and flavor and overall similarity of butters spiked with BMS to off-flavored butter.

compound(s) plays an important role in flavor (Guth and Grosch, 1994; Schieberle and Hofmann, 1997) as both concentration in the food and sensory threshold are simultaneously taken into account. Model addition and subtraction studies, also called n-1 or omission studies, involve sequential addition or subtraction of compounds of interest to or from the food or model matrix followed by sensory analysis after each step. This approach has been used to characterize key components of complex flavors such as strawberry juice (Schieberle and Hofmann, 1997) and fresh milk powder (Karagul–Yuceer et al., 2004). Table 2.3 shows similarity scores for n-1 model system mixtures of key components of fresh rehydrated low-heat skim milk powder (SMP). Finally, discussion of model system analysis is not complete without addressing selection of the model matrix. In the case of off-flavor identification, the actual food can often be used for direct addition of compounds. In other cases, such as when the goal is to identify key components of fresh or desirable flavor, selection of a matrix is more troublesome. Polarity, ionic strength, and pH can all impact thresholds and thus flavor. Further, the role of other components at below-threshold concentration may also play a role in flavor perception. If a matrix other than the actual food is to be used, it should be as similar as possible in composition and properties to the food being studied in order to make powerful conclusions regarding the role of specific components. Table 2.4 demonstrates the differences in sensory thresholds between water and fluid skim milk. Note the threshold differences between these two very similar matrices.

Table 2.3 Average Aroma Similarity Scores of Skim Milk Powder Models with One Compound Removed to Fresh Rehydrated Skim Milk Powder

n-1 Model	Average similarity
Complete	9.0a
Furaneol	7.9b
Butanoic acid	6.5cd
Methional	6.8c
o-Aminoacetophenone	8.8a
δ-Decalactone	8.0b
Vanillin	6.0d
Hexanoic acid	6.8cd
2-Acetyl-2-thiazoline	6.5c
Octanoic acid	7.8b
γ-Dodecalactone	7.5b
(E,E)-2,4-decadienal	7.5b
Maltol	6.8c
Dimethylsulfide	6.9c

Note: Similarity scores followed by different letters are different (p < 0.05)

Source: From Karagül-Yüceer, Y., Drake, M.A., and Cadwallader, K.R., 2004. Evaluation of the character impact odorants in skim milk powder by sensory studies on model mixtures. *J. Sensory Stud.* 19: 1–14. With permission.

Table 2.4 Best Estimate Values of Threshold (BET) for Selected
Compounds in Water and in Fluid Skim Milk

Odorant	BET (µg/l)	
	In water	In skim milk
Furaneol	39 ± 24	575 ± 45
Butanoic acid	1274 ± 62	3090 ± 10
Methional	0.4 ± 0.08	5 ± 9
o-AAP	0.3 ± 0.09	84 ± 33
δ-Decalactone	30 ± 10	603 ± 11
Vanillin	64 ± 47	7413 ± 14
Hexanoic acid	36 ± 8	1273 ± 11
2-Acetyl-2-thiazoline	1 ± 0.2	2 ± 1
Octanoic acid	1405 ± 470	44668 ± 15
γ-Dodecalactone	5 ± 3	933 ± 8
(*E,E*)-2,4-decadienal	0.2 ± 0.09	1 ± 2
Maltol	210 ± 50	16596 ± 16
Dimethylsulfide	3 ± 2	8 ± 3

Source: Adapted from Karagül-Yüceer, Y., Drake, M.A., and Cadwallader, K.R., 2004.
Evaluation of the character impact odorants in skim milk powder by sensory
studies on model mixtures. *J. Sensory Stud.* 19: 1–14.

III.   Analysis of multiple samples or multiple volatile components: another approach

In many situations, identification of a specific flavor is not the goal. Instead, enhanced understanding of flavor components, flavor formation or stability is desired. One or several compounds may be involved, or the exact number or type(s) of compound(s) responsible is not known. Further, a large number of samples may need to be evaluated. In these cases, the three-step approach described previously is simply not feasible. In these cases, however, the two basic components, instrumental analysis and sensory analysis are still crucial requirements. The primary difference is that threshold analysis and model system analysis is not conducted. GC-O may also not be conducted due to its time-consuming nature. Karagul-Yuceer et al. (2001, 2002) conducted studies on flavor contributors in both fresh and stored SMPs. SMPs were selected and profiled by trained sensory panelists. Volatile odor-active components in SMP were then identified and characterized. Specific sources for particular flavors were not identified. However, general groups of compounds contributing to SMP flavor were identified, and a comparison of results between fresh and stored SMP leads to some general conclusions about what types of sensory and volatile compound changes can occur between fresh and stored SMP. Carunchia-Whetstine and Drake (2006) built on that premise of SMP flavor to follow sensory and volatile compound changes in the same SMP and whole milk powder (WMP) across 36 months of storage. Similarly, Karagul-Yuceer et al. (2003) and Carunchia Whetstine et al. (2003) investigated sensory flavor and instrumental volatile components of liquid Cheddar whey. Frozen wheys were collected from different manufacturers and starter culture rotations.

Sensory profiling and instrumental volatile analyses were conducted. The results demonstrated sensory flavor and instrumental volatile variability with manufacture facility and starter culture rotation. Such results provide a frame of reference for more recent research on the flavor variability (sensory and instrumental) of whey proteins (Caruncia Whetstine et al., 2005). Two examples of this more general flavor research approach with SMP follow.

A. Flavor formation across SMP production

Large quantities of SMP products are produced in the U.S., many of which are directed toward specific end-product applications due to specific heat treatment throughout the drying process. The quality of the raw milk is certainly a determining factor in SMP flavor, but the processing steps that produce SMP can also affect the flavor. The heat treatment of the fluid milk before it enters the spray-drier is the criterion used to determine the SMP category. The heat treatments are 72°C/15 sec, 82°C/3 min and 82°C/30 min for low, medium, and high heat, respectively (Rehman et al., 2003). The heat treatment affects the functionality of the SMP by denaturing whey proteins. The extent of denaturation is important to the use of the end product because of flavor and functionality differences (Walstra et al., 1999a). Although low-heat SMP is the most commonly produced type of SMP, medium-heat SMP production is increasing. Medium-heat SMP is used in products such as milk-based drinks and ice cream, and it is also the primary SMP utilized in the global market (Varnum and Sutherland, 1994). High- and medium-heat SMP characteristically have a noticeably higher-cooked flavor compared to low-heat SMP (Karagul-Yuceer et al., 2001). Inappropriate, or inaccurate heat treatments, or holding times at any stage of the drying process, can cause undesirable flavors or inconsistencies in SMP flavor of any heat type.

The total time required to convert raw milk to SMP is less than 1 h. Large commercial facilities may process milk continuously for 40 h followed by a brief shut down for cleaning and sanitation purposes. Flavor (desirable as well as undesirable) formation can occur at any stage in SMP production. Caudle et al. (2005) demonstrated that off-flavors in low-heat SMP were carried through into a variety of ingredient applications in which they negatively impacted consumer acceptance. Clearly, sensory quality of SMP is crucial. In this study, we examined the flavor formation of medium-heat SMP throughout the production process, with emphasis on the steps in which heat is applied to the product.

1.   Materials and methods

Samples. Raw milk (12% solids, 9% solids not fat [SNF], balance tank (9% SNF), HTST (high temperature, short time) pasteurized (9% SNF), evaporated milk (50% SNF) and SMP (94% SNF) samples, were collected from a large west coast production facility on three different occasions in fall 2004 (Turlock, CA). Samples (4 l liquid / 400 g powder, 5 sampling points in triplicate) were collected across processing facility run-time (beginning ~2 h, middle ~20 h, end ~38 h), packed in Mylar pouches (Impak, Los Angeles, CA), heat-sealed, and frozen at −20°C. Frozen products were packed on dry ice and shipped by overnight carrier to North Carolina State University. Upon receipt, pouches were examined for damage and subsequently were stored at −20°C in the dark until further analysis. Proximate analysis (ash, fat, protein, total solids, and microbial quality) was conducted in duplicate on each product.

Descriptive analysis. Each of the products were evaluated using the descriptive sensory language and methods described by Drake et al. (2003) (Table 2.5). The evaporated milk and SMP were reconstituted (10% SNF) using deodorized water. Evaporated product was also evaluated directly (undiluted). Fluid samples (30 ml) and undiluted evaporated milk (5 g) were dispensed into 2 oz soufflé cups with lids and 3-digit codes. Sensory analysis was conducted at 12°C. Six trained panelists (5 female, 1 male) evaluated each sample in duplicate. Each panelist had more than 50 h of training on the descriptive sensory analysis of dried dairy ingredients with a previously identified descriptive language and the Spectrum™ descriptive analysis method (Drake et al., 2003; Meilgaard et al., 1999).

Direct solvent extraction. Volatile components were extracted from each sample in a similar manner to that described previously for butter. The volume of total solids, and the ether-to-volume ratio was standardized for solvent extraction. The fluid samples required more fluid volume, as well as a greater volume of ether, compared to rehydrated samples (evaporated milk, SMP). Evaporated milk and SMP were rehydrated (10% solids) using deodorized water (250 ml). Fluid milks (raw, balance tank, and HTST pasteurized, 522 ml) and rehydrated milks (evaporated and SMP, 250 ml) were placed in clear Pyrex glass beakers (Corning, Acton, MA). Ten microliters of internal standard (0.185 ppm of 2-methyl-3-heptanone + 2-methyl valeric acid in methanol, Aldrich Chem. Co., Milwaukee, WI) were added to each sample, and the sample pH was measured with an Orion 250A+ pH meter (Thermo Electron Corp., Waltham, MA). Samples (65 to 75 ml) were dispersed into either 4 or 6 Teflon-coated-centrifuge bottles (Nalgene, Rochester, NY). Ether (75 ml for fluid milk, 50 ml for rehydrated samples) was added to each of the centrifuge bottles. The bottles were agitated for 30 min before they were centrifuged at 735 g for 15 min. This process was repeated twice, first with the addition of 50 ml of ether and again with the addition of 25 ml of ether for all samples. The bottle was subsequently centrifuged without addition of ether at 735 g for 20 min. After each centrifugation, the ether was collected from each bottle and pooled into a Quorpak clear wide-mouth jar (Fisher, Pittsburgh, PA) and stored at −20°C. Using a Vigreux column, each sample was concentrated to 50 ml in a round-bottom flask at 40°C in a water bath (Isotemp 110, Fisher, Pittsburgh, PA), transferred to small Quorpak jars (250 ml. Fisher, Pittsburgh, PA), and stored at −20°C until SAFE.

Table 2.5 Descriptive Sensory Language for Fluid Milk and Rehydrated SMP

Descriptor	Definition	Reference	Preparation
Cooked/ sulfurous	Aromatic associated with cooked milk	Heated milk	Heat pasteurized skim milk to 85°C for 45 min
Sweet aromatic/ cake mix	Sweet aromatic associated with dairy products	Pillsbury white cake mix Vanillin	Dilute 5 mg of vanillin in skim milk
Cardboard/ wet brown paper	Aromatic associated with cardboard	Wet brown paper or cardboard	Soak 2-cm square of brown paper bag or cardboard in warm water for 30 min
Fatty/fried	Aromatic associated with stale oils or old frier oil	2, 4 decadienal	10 ppb in skim milk
Vitamin	Aromatic associated with vitamin supplements or rubber	Enfamil–liquid Polyvisol vitamins	
Free fatty acid	Aromatic associated with free fatty acids	Feta cheese or butyric acid	Crumbled Feta cheese or 20 ppm butyric acid in skim milk
Metallic/ serum-like	Aromatic associated with rare steak juice	Rare steak juice	
Sweet taste	Basic taste associated with sugars	Sucrose	5% Sucrose solution
Salty	Basic taste associated with salts	NaCl	2% NaCl solution
Sour	Basic taste associated with acids	Citric acid	1% Citric acid solution
Bitter	Basic taste associated with various compounds	Caffeine	0.5% Caffeine solution
Astringent	Drying or puckering of oral tissues	Tea	Soak 6 tea bags in water for 10 min

Source: Adapted from Drake, M.A., Karagul-Yuceer, Y., Cadwallader, K.R., Civille, G.V., and Tong, P.S. 2003. Determination of the sensory attributes of dried milk powders and dairy ingredients. *J. Sensory Stud.* 18: 199–216; Carunchia-Whetstine et al., 2003.

Solvent-assisted flavor evaporation (SAFE). Volatile compounds were distilled from solvent extracts using SAFE, followed by phase separation and concentration as described for butter. Extracts were placed into small screw cap vials (2 ml) and stored at −20°C.

Gas chromatography-olfactometry (GC-O) and AEDA. GC-O and AEDA were used to characterize aroma-active compounds contributing to flavor in SMP samples as described for butters with a few modifications. The oven temperature was initially 40°C for 3 min. and then it was 15°C/min to a final temperature of 200°C for 15 min.

Gas chromatography-mass spectrometry (GC-MS). GC-MS was conducted on each concentrated sample fraction as described for butter.

Statistical analysis. Analysis of variance with means separation (least-squares mean) was conducted to determine differences between samples and differences between sample collection time (beginning, middle, or end of production run). Both main effects (sample, collection time) as well as interactions (sample collection time) were addressed. Statistical analysis was performed using SAS version 8.2 (Cary, NC).

2. Results

Sensory analysis. The sensory attributes of the milk changed significantly throughout the SMP production process ($p < .05$) (Table 2.6).

Table 2.6 Sensory Attributes of SMP throughout the Production Process[a]

Attribute	Raw milk	Bulk tank	HTST	Evaporated	Evaporated (9% SNF)	SMP
Cooked	ND	1.92	3.50	6.00	1.83	4.25
Caramelized	ND	ND	ND	3.67	1.08	1.58
Milk fat	2.50	0.92	0.52	1.25	ND	ND
Sweet aromatic	1.42	2.28	2.77	1.50	1.08	2.93
Sweet taste	2.00	2.67	2.95	3.55	1.41	2.55
Salty taste	0.58	0.58	ND	4.00	1.50	0.58
Feed	1.42	1.33	1.42	ND	ND	ND
Free fatty acid[e]	0.65	ND	ND	ND	ND	ND
Vitamin	ND	ND	ND	1.58	0.50	ND
Metallic/serum	1.67	0.67	ND	ND	ND	ND
Astringency	1.08	1.25	1.67	1.75	2.25	2.46

Intensities are scored on a 15-point scale where 0 = none and 15 = very high (Data from Drake, M.A., Karagul-Yuceer, Y., Cadwallader, K.R., Civille, G.V., and Tong, P.S. 2003. Determination of the sensory attributes of dried milk powders and dairy ingredients. *J. Sensory Stud.* 18: 199–216; Meilgaard, M.C., Civille, G.V., and Carr, B.T., 1999. *Sensory Evaluation Techniques,* 3rd ed., CRC Press: Boca Raton, FL.)

Means in a row followed by different letters are different ($p < 0.05$).

ND = not detected

There was not a collection time effect ($p > 0.05$).

There was a collection time effect for this attribute for raw milk collected at the beginning of production.

Samples from different times within the production run (beginning, middle, and end) were not different in sensory profiles ($p < 0.05$) (data not shown). The flavors of fluid milk and SMP are delicate, and thus intensities are low, consistent with previous studies (Karagul-Yuceer et al., 2001, 2002; Drake et al., 2003). Milk fat flavor was high in the raw milk and decreased at subsequent stages owing to skimming before entering the bulk tank. Metallic or serum and feed flavors were found in the raw and bulk tank milks, and decreased to nondetectable levels with increasing heat treatment. These are flavors typically associated with raw milk. Cooked flavor increased as the heat treatment increased. The sweet aromatic flavor increased throughout the process, with the exception of the evaporated sample. SMP is typically much higher in sweet aromatic flavor compared to fluid milk (Drake et al., 2003), and this is likely due to caramelization compounds produced during heat treatment (Karagul-Yuceer et al., 2004). The volatile flavors, except sweet aromatic, were consistently higher in the undiluted evaporated sample than in other samples. There was collection time effect for any processed (heat-treated) sample ($p > 0.05$) suggesting that equipment burn-on, or soiling throughout 40 h of production, does not impact sensory properties of the finished product. However, low levels (intensity = 0.75) of free fatty acid were detected in the raw milk from the first collection, indicative of hydrolytic rancidity. Free fatty acid flavors were not detected in subsequent processed samples from this collection point.

Instrumental analysis. Thirty-nine aroma-active compounds were identified in milk samples throughout the SMP production process (Table 2.7 and Table 2.8). Ethyl hexanoate was found in raw milk from the end-of-production samples. This compound was previously identified as one of the most potent odorants in raw cow's milk (Friedrich and Acree, 1998). Butanoic acid (cheesy) and hexanoic acid (sour or sweaty) were found in samples from each stage throughout production. Other free fatty acids were found only in samples at the middle- and end-of-the-production run. These acids were pentanoic (sour or sweaty) and 2/3-methyl butanoic (cheesy or dried fruit) acids. Karagul-Yuceer et al. (2001) documented these free fatty acids, caused by fat hydrolysis, as major contributors to SMP flavor. Other flavors identified by post peak intensity were heat-generated compounds, including methional, 2-acetyl-1-pyrroline and homofuraneol, which have potato, popcorn-like, and burnt-sugar aromas, respectively. All of these compounds have been documented as thermally induced compounds in SMP (Karagul-Yuceer et al., 2001). Hexanal and γ-decalactone were also found in some of the heated samples. Both compounds were noted as potent aroma-active compounds in heated milk (Friedrich and Acree, 1998). The intensities of these compounds increased from the raw milk to the SMP as the heat treatment increased. Few changes were seen throughout the production run from

Table 2.7 Summary of Aroma-Active Compounds Identified in SMP throughout Processing (Beginning of Run)

| Number | Compound | Fraction | Odor at the sniffer port[a] | RI[b] | | Post peak intensity at the sniffer port (n = 4) (Log$_3$ Dilution factor)[c] | | | | | Method of identification[a-d] |
				DB-5MS	DB-WAX	Raw milk	Bulk tank	HTST	Evap-orated milk	SMP	
1	Unknown	NB	Plastic/solventy	788		1.5	ND	ND	1.6	1.3(<1)	RI, odor
2	Hexanal	NB	Green/sweet	823		ND	ND	ND	1.5	4.0	RI, odor, MS
3	Unknown	NB	Skunk	836		ND	2.0	ND	ND	ND	
4	Butanoic acid	Ac	Rancid/Cheese	858	1603	2.2	ND	1.8	1.5	2.5(5)	RI, odor, MS
5	Propionic acid	Ac	Cheesy/sour	865	1523	3.7	3.2	2.0	1.2	3.0(5)	RI, odor, MS
6	Unknown	NB	Fruity	866		1.3	ND	1.17	ND	2.3(<1)	
7	Methional	NB	Potato	933		ND	3.0	2	2	3(3)	RI, odor
8	2-Acetyl-1-pyrroline	NB	Popcorn	942		1.5	1.8	3.0	2.25	3.1(2)	RI, odor
9	DMTS	NB	Cabbage	987		ND	1.0	2.5	2	2.6(3)	RI, odor, MS
10	1-Octen-3-one	NB	Mushroom/metallic	995		1.9	1.8	1.0	1.9	2.5(1)	RI, odor, MS
11	Hexanoic acid	Ac	Sour/sweaty	1013	1852	1.8	ND	1.5	1.5	1.5(1)	RI, odor, MS
12	Octanal	NB	Fruity/citrus	1020		1.5	2.2	ND	1.6	2.5(<1)	RI, odor, MS
13	(E)-2-Octenal	NB	Stale/musty	1084		1.8	1.5	1.2	1.5	ND	RI, odor, MS
14	Nonanal	NB	Earthy/fatty	1102		ND	2.0	ND	ND	ND	RI, odor, MS

(continued)

Table 2.7 (Continued) Summary of Aroma-Active Compounds Identified in SMP throughout Processing (Beginning of Run)

Number	Compound	Fraction	Odor at the sniffer port[a]	RI[b] DB-5MS	RI[b] DB-WAX	Post peak intensity at the sniffer port (n = 4) (Log₃ Dilution factor)[c] Raw milk	Bulk tank	HTST	Evaporated milk	SMP	Method of identification[a-d]
15	Unknown	NB	Earthy/burnt	1110		ND	ND	2.3	2.8	2.3(<1)	
16	Homofuraneol	Ac	Burnt sugar	1121	2045	ND	ND	ND	1.3	3.0(3)	RI, odor, MS
17	3,6-Nonadienal	NB	Fatty	1123		2.0	2.0	2.0	2.2	2.8(<1)	RI, odor, MS
18	Unknown	NB	Oxidized/roasted	1137		ND	ND	2.0	2.2	2.5(<1)	
19	(E,Z)-2,6-Nonadienal	NB	Rosy/cucumber	1169		1.8	1.6	1.5	2.0	2.0(3)	RI, odor, MS
20	Butyl hexanoate	NB	Cucumber/floral	1204		ND	ND	ND	ND	3.0(3)	RI, odor, MS
21	Decenal	NB	Old books	1229		ND	2.0	ND	ND	1.5(1)	RI, odor, MS
22	Unknown	NB	Rosy/waxy	1341		1.5	1.5	1.8	ND	ND	
23	γ-Nonalactone	NB	Sweet/honey	1416		ND	ND	ND	ND	1.2(2)	RI, odor
24	γ-Decalactone	NB	Sweet/fruity	1524		1.2	ND	ND	ND	ND	RI, odor, MS
25	γ-Dodecalactone	NB	Peach	1695		1.5	1.5	2.0	ND	ND	RI, odor, MS
26	δ-Dodecalactone	NB	Peach	1736		1.0	1.0	2.5	ND	1.0(<1)	RI, odor, MS

[a] Odor description at the GC-sniffing port.

[b] Retention indices were calculated from GC-O data.

[c] Flavor dilution factors were determined on a DB-5 column for NB compounds, and on a DB-Wax column for Ac compounds.

[d] Compounds were identified by comparison with the authentic standards on the following criteria: retention index (RI) on DB-Wax and DB-5MS columns, odor property at the GC-sniffing port, and mass spectra in the electron impact mode.

Table 2.8 Summary of Aroma-Active Compounds Identified in SMP throughout Processing (End of Run)

Number	Compound	Fraction	Odor at the sniffer port[a]	RI[b]		Post peak intensity at the sniffer port (n = 4) (Log₃ Dilution factor)[c]					Method of identification[a-d]
				DB-5MS	DB-WAX	Raw milk	Bulk tank	HTST	Evaporated milk	SMP	
1	Diacetyl	NB	Buttery	640		ND	ND	1.5(<1)	ND	ND	RI, odor, MS
2	2/3 Methyl butanal	NB	Chocolate/malty	686		1.5(<1)	ND	ND	ND	ND	RI, odor, MS
3	Unknown	NB	Plastic/solventy	794		1.5(<1)	1.2	1(<1)	ND	1.3(<1)	RI, odor
4	Ethyl butyrate	NB	fruity/sweet	815		2.5(3)	1.7	1.8(<1)	1.5(<1)	ND	RI, odor, MS
5	2-Methyl-3-furanthiol	NB	Brothy/dirty	830		ND	ND	1.5(<1)	ND	3.0(<1)	RI, odor
6	Butanoic acid	Ac	Rancid Cheese	858	1615	4.2(4)	3.5	3.3(2)	3.0(2)	3.1(3)	RI, odor, MS
7	2/3-Methyl butanoic acid	Ac	Cheesy/dried fruit	930	1688	3.5(3)	4.1	4.0(3)	1.5(3)	ND	RI, odor, MS
8	Pentanoic Acid	Ac	Sweaty/waxy	930	1758	ND	ND	ND	3.6(<1)	3.3(<1)	RI, odor, MS
9	Propionic acid	Ac	Cheesy/sour	865	1477	3.7(<1)	3.2	2.0(<1)	1.2(<1)	2.5(<1)	RI, odor, MS
10	Unknown	NB	Fruity	866		3.5(<1)	ND	1.5(<1)	ND	ND	RI, odor, MS
11	Methional	NB	Potato	933		3.5(1)	2.2	2.8(4)	2.1(3)	2.7(1)	RI, odor
12	2-Acetyl-1-pyroline	NB	Popcorn	942		ND	2.0	2.2(3)	1.8(<1)	2.1(1)	RI, odor
13	Ethyl hexanoate	NB	Fruity	972		2.0(<1)	ND	ND	ND	ND	RI, odor, MS
14	Dimethyl trisulfide	NB	Cabbage	987		3.5(<1)	2.6	3.1(4)	2.5(<1)	3.0(1)	RI, odor, MS
15	1-Octen-3-one	NB	Mushroom/metallic	995		1.5(<1)	ND	2.0(<1)	2.0(<1)	1.2(<1)	RI, odor, MS
16	Hexanoic acid	Ac	Sour/sweaty	1013	1830	ND	ND	2.5(<1)	4.5(<1)	2 (3)	RI, odor, MS
17	Octanal	NB	Fruity/citrus	1020		1.7(<1)	1.5	2.0(<1)	2.0(<1)	2.0(<1)	RI, odor, MS
18	2-Acetyl thiazole	NB	Popcorn	1038		ND	ND	2.0(<1)	ND	2.0(<1)	RI, odor, MS
19	(E)-2-Octenal	NB	Stale/musty/fatty	1082		1.5(1)	1.0	ND	1.5(<1)	2.0(<1)	RI, odor, MS
20	2-Acetyl-2-thiazoline	NB	Popcorn/sweet	1100		ND	ND	ND	ND	3.0(1)	RI, odor, MS
21	Nonanal	NB	Earthy/fatty	1102		1.5(3)	1.5	2.5(3)	ND	ND	RI, odor, MS
22	2-Phenethanol	ND	Rosey/cuke	1117		2.2(1)	3.5	ND	ND	2.5(<1)	RI, odor, MS
23	Homofuraneol	Ac	Burnt sugar	1121	2065	ND	ND	1.5(1)	1.5(<1)	1.5(1)	RI, odor, MS

(continued)

Table 2.8 (Continued) Summary of Aroma-Active Compounds Identified in SMP throughout Processing (End of Run)

Number	Compound	Fraction	Odor at the sniffer port[a]	RI[b] DB-5MS	RI[b] DB-WAX	Post peak intensity at the sniffer port (n = 4) (Log₃ Dilution factor)[c] Raw milk	Bulk tank	HTST	Evaporated milk	SMP	Method of identification[a-d]
24	3,6-Nonadienal	NB	Fatty	1133		ND	1.0	2.5(<1)	2.5(<1)	2.8(<1)	RI, odor, MS
25	Unknown	NB	Oxidized/roasted	1137		ND	ND	2.0(1)	2.2(<1)	2.5(<1)	RI, odor, MS
26	o-Cresol	NB	Cowy	1154		ND	ND	ND	ND	2.0(1)	RI, odor, MS
27	(E,Z)-2,6-Nonadienal	NB	Rosy/cucumber	1169		2.0(1)	2.5	2.0(<1)	2.5(<1)	2.0(1)	RI, odor, MS
28	(E)-2-Nonenal	NB	Old books	1186		2.2(<1)	1.9	2.4(<1)	1.8(<1)	2.5(<1)	RI, odor, MS
29	2-Isobutyl-3-methoxy pyrazine	NB	Bell pepper	1193		ND	ND	2.5(<1)	1.5(<1)	2.2(1)	RI, odor, MS
30	Unknown	NB	Cucumber/floral	1204		ND	2.0	ND	ND	ND	RI, odor
31	(E,E)-2,4-Nonadienal	NB	Earthy/sweet/nutty	1220		1.5(1)	2.0	2.0(<1)	ND	ND	RI, odor, MS
32	2-Decenal	NB	Old books	1229		ND	1.5	ND	ND	3.5(<1)	RI, odor, MS
33	Unknown	NB	Rosy/waxy	1341		1.5(1)	1.5	1.8(<1)	ND	ND	RI, odor
34	Skatole	NB	Fecal	1501		2.5(<1)	ND	2.5(1)	1.7(<1)	1.8(1)	RI, odor
35	γ-Decalactone	NB	Sweet/fruity	1524		2.5(<1)	2.5	2.25(<1)	ND	ND	RI, odor, MS
36	γ-Octalactone	NB	Coconut	1563		3.0(<1)	ND	ND	1.7(<1)	ND	RI, odor, MS
37	6-(Z)-Dodecen-γ-lactone	NB	Peach/waxy/coconut	1609		ND	1.8	2.5(<1)	2.0(<1)	1.3(1)	RI, odor, MS
38	γ-Dodecalactone	NB	Peach	1695		ND	2.5	2.7(3)	1.5(<1)	1.2(<1)	RI, odor, MS
39	δ-Dodecalactone	NB	Peach	1736		2.0(3)	2.2	2.0(<1)	ND	2.0(<1)	RI, odor, MS

[a] Odor description at the GC-sniffing port.

[b] Retention indices were calculated from GC-O data.

[c] Flavor dilution factors were determined on a DB-5 column for NB compounds, and on a DB-Wax column for Ac compounds.

[d] Compounds were identified by comparison with the authentic standards on the following criteria: retention index (RI) on DB-Wax and DB-5MS columns, odor property at the GC-sniffing port, and mass spectra in the electron impact mode.

beginning to end. The most prominent difference between the time-points was that there were more compounds identified in samples from the end of the run as compared to beginning and middle samples. Compounds such as o-cresol (cowy), (E,E)-2,4-nonadienal (earthy/sweet/nutty), 2-methyl-3-furanthiol (brothy or dirty) and 2-acetyl-2-thiazoline (popcorn or sweet) were due to burn-on residues and increased at the end-of-the-production run.

AEDA was performed on samples from the end-of-the-production run based on the post peak intensity results as there were a larger number of compounds, representative of all three time points found in these samples. The most prominent changes in volatile compounds occurred at the major heating steps, HTST pasteurization, evaporation, and spray-drying. Therefore, we conducted AEDA on the raw milk, HTST pasteurized milk, evaporated milk, and SMP (Table 2.8). Previous research by Karagul–Yuceer et al. (2001) found that heat-induced compounds, such as methional and 2-acetyl-1-pyrroline, as well as free fatty acids (butanoic acid and pentanoic acid), caused by hydrolysis of milk-fats, are important odorants in SMP. The same study also suggested that the number of compounds, as well as their concentrations, contributes largely to SMP flavor. Butanoic acid had a higher $\log_3$ FD value in samples from the beginning of the run compared to the middle and end of the run. Propionic acid also had a high dilution factor in the beginning sample ($\log_3$ FD value = 5), but this is less than 1 in subsequent samples. Methional and 2-acetyl-1-pyrroline had a lower $\log_3$ FD value in the end-of-run samples than the other samples. These results suggest that fewer, more potent compounds make up the flavor of the beginning and middle samples, but more compounds (less potent) combine to make up the flavor of the end-of-production sample. Previous research has shown that raw milk has fewer common aroma-active compounds present (7) than heated milk (15) and that the potency of the compounds was different between the samples (Friedrich and Acree, 1998). Sensory differences were not detected between the three SMP collected at different times in production. However, differences in volatile composition may contribute to differences in flavor stability, and should be addressed in future work.

In this study, the combination of sensory and instrumental analyses confirms the role of heat-generated compounds in the flavor of SMP, and documents concurrent sensory and instrumental changes in fluid milk across SMP production and across production run time. The study also demonstrates the inherent danger of instrumental analysis alone — even when previous studies have characterized sources of sensory flavors. In the current study, differences in volatile components in SMP were observed by GC-O and by quantification. However, these differences had no perceived effect on the sensory perception of flavor of the rehydrated SMP.

B.   *Flavor changes in SMP with storage*

Fresh SMP should ideally exhibit a mild and bland flavor reminiscent of fluid skim milk (Bodyfelt et al., 1988). The proposed shelf life of SMP varies.

A shelf life of anywhere between 6 to 36 months for noninstantized unfortified SMP under optimal storage conditions has been proposed by various sources (Varnum and Sutherland, 1994; USDEC, 2001; ADPI, 2002). Characterizing SMP flavor variability, flavor stability and its role in consumer acceptance is crucial. Further, identification of key compounds that are associated with flavor degradation is important in order to identify rapid instrumental methods for evaluation of SMP flavor quality. Previous studies have evaluated sensory properties concurrently with labor-intensive solvent extraction instrumental approaches (Karagul–Yuceer et al., 2001, 2002; Carunchia-Whetstine and Drake, 2006). The objective of the current study was to evaluate the efficacy of a more rapid instrumental volatile analysis to assess changes in SMP across storage time. SMP were evaluated by descriptive sensory analysis; solid-phase microextraction (SPME) was used for volatile extraction.

1. Materials and methods

Skim milk powders (SMP). Low-heat SMP commercially packaged in 3-ply 22-kg bags were received from six commercial facilities on the west coast of the U.S. within 3 weeks of production. SMP were stored in the dark at 21°C, 60% RH (relative humidity). Upon sampling, the bags were cut open and the top 4 cm of exposed surface was removed. A 500 g sample was then collected for analysis.

Descriptive analysis. Products were rehydrated and evaluated using the methods described previously for SMP (Table 2.5).

Solid-phase microextraction (SPME). SMP were rehydrated to 10% solids using deodorized water. Each sample (100 ml) was transferred into prewashed Quorpak clear standard wide mouth jars (250 ml; Fisher, Pittsburgh, PA), followed by the addition of 2.5 µl of internal standard (9.07 ppm 2-methyl-3-heptanone in methanol; Aldrich Chem. Co., Milwaukee, WI). The sample (20 g) was transferred into precleaned 40 ml amber vials with screw caps and Teflon silica septa (Supelco, Bellefonte, PA). A magnetic octagonal stirring bar (8 mm OD × 13 mm length; Fisher, Pittsburgh, PA) and commercial sodium chloride (1 g) were added. Each sample was allowed to equilibrate in the vial at 40°C for 30 min using a heating/stirring module (Reacti-Therm™, Pierce Biotechnology, Inc., Rockford, IL). Volatile odor compounds were extracted by exposing a 2 cm-50/30 µm divinylbenzene/carboxen/polydimethylsiloxane (DVB/CAR/PDMS) StableFlex SPME fiber (Supelco, Bellefonte, PA) to the headspace of each sample at 40°C for 30 min. The fiber was desorbed at 250°C for 5 min in the injection port fitted with a SPME inlet liner (Supelco, Bellefonte, PA), at a depth of 3 cm on a Varian CP–3380 GC/Saturn 2000 ion-trap-mass selective detector (Varian Inc., Palo Alto, CA). Separations were performed on a fused-silica capillary column (Rtx-5 30 m length × 0.25 mm ID × 0.25 µm df; Restek, Bellefonte, PA). Helium gas was used as a carrier at a constant flow of 1 ml/min. The oven temperature was programmed from 40 to 250°C at a rate of 8°C/min with initial and final hold times of 5 min. MSD conditions were as follows: transfer line 120°C, manifold 80°C, ion trap 150°C, ionization via automatic electronic ionization, mass range m/z 35 to 350, EM voltage 2135 V; scan/sec = 1. Each sample was analyzed in triplicate. Volatile compounds were identified using the NIST05 mass spectral library, and retention indices of authentic standard compounds analyzed under identical conditions. The compounds were selected for quantification (relative abundance) based on a comparison of the chromatograms of fresh time zero SMP and 30-month-old SMP and examination of previous literature.

Statistical analysis. Principal component analysis using the correlation matrix was conducted on the sensory and instrumental data sets individually to compare how the SMP were differentiated by each approach (SAS version 9.1, Cary, NC).

2. Results

The SMP were differentiated by both sensory and instrumental volatile analyses (Figure 2.4 to Figure 2.7). By sensory analysis, the 30-month SMP

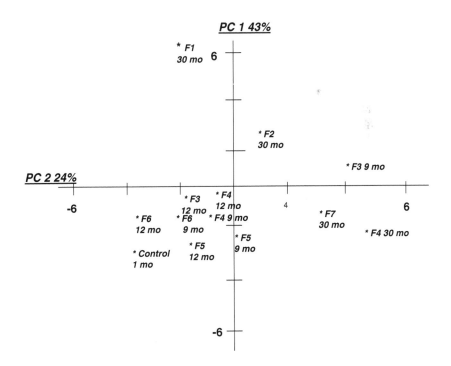

Figure 2.4 Principal component biplot of sensory analysis of SMP from various facilities stored for various times.

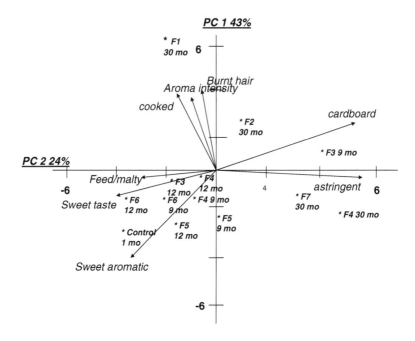

Figure 2.5 Principal component biplot of sensory analysis of SMP from various facilities stored for various times. Attributes are overlaid as vectors.

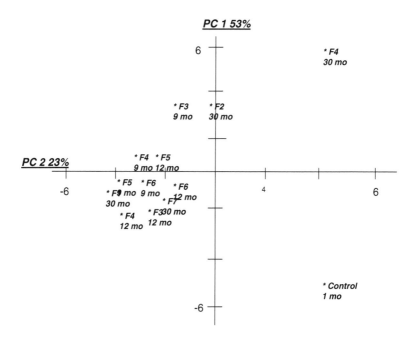

Figure 2.6 Principal component biplot of instrumental profiles of SMP from various facilities stored for various times.

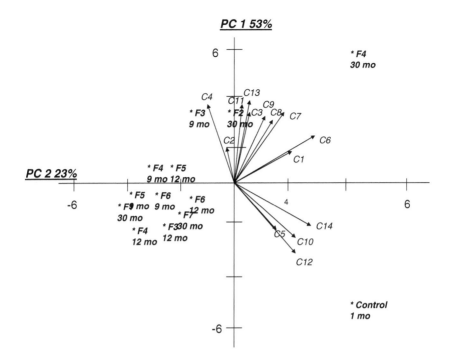

Figure 2.7 Principal component biplot of Instrumental profiles of SMP from various facilities stored for various times. Compounds (Table 2.9) are overlaid as vectors.

samples and one 9-month SMP sample were most distinct from the control or fresh SMP (Figure 2.4) samples. As SMP aged, cardboard flavor and astringency increased. Further, as SMP aged (9-month and some 12-month SMP), sensory flavor changes were initially characterized by a decrease in "fresh" flavors (decreases in sweet aromatic, cooked, and sweet taste) rather than the presence of cardboard flavor. These results are consistent with previous sensory studies on SMP (Drake et al., 2003; Carunchia-Whetstine and Drake, 2006). The SMP F1 30 month was differentiated by sensory analysis from the rest of the SMP (Figure 2.4, Figure 2.5). This SMP was characterized by a high-aroma intensity (sulfur, eggy), high cooked flavor, and a low intensity of burnt-hair or burnt-feather flavor. These sensory attributes made it quite distinct from the other SMP. This SMP was a medium-heat SMP, whereas the other SMP were low-heat. Medium-heat SMP are characterized by more sulfurous aromas and higher cooked flavors compared to low-heat SMP owing to the higher heat profiles they are subjected to during processing (Karagul-Yuceer et al., 2002). The presence of the low intensity of burnt-hair or burnt-feather flavor is an off-flavor that can develop with storage time in medium-heat SMP.

    SMP were also differentiated by instrumental analysis (Figure 2.6, Figure 2.7). As with sensory analysis, as the SMP aged, they became differentiated from the control fresh SMP. As the SMP aged, the relative abundance of many aldehydes (Table 2.9) increased, whereas the abundance of maltol decreased.

Table 2.9 Mean Relative Abundance for Selected Compounds Identified by SPME for Fresh vs. Stored SMP

Relative abundance (Standard deviation)	+ Treatment	Control 1 mo 1	F1 30 mo 2	F1 30 mo 3	F2 30 mo 4	F4 30 mo 5	F4 12 mo 6	F5 12 mo 7	F6 12 mo 8	F3 12 mo 9	F6 9 mo 10	F5 9 mo 11	F4 9 mo 12	F3 9 mo 13
Furan, 2,5-dimethyl-	Compound 1	12.49 (15.81)	11.44 (1.91)	8.37 (7.93)	24.99 (4.55)	49.74 (7.70)	8.09 (2.92)	13.18 (1.63)	34.01 (1.29)	4.45 (3.98)	8.84 (7.72)	7.74 (6.75)	7.67 (5.38)	9.92 (9.08)
Disulfide, dimethyl	Compound 2	0.00 (0.00)	0.98 (1.11)	8.99 (1.75)	2.01 (0.41)	3.55 (0.88)	0.41 (0.07)	ND	0.95 (0.11)	0.06 (0.11)	ND	ND	0.45 (0.39)	ND
Hexane, 2,4-dimethyl-	Compound 3	1.61 (1.16)	6.95 (2.02)	9.62 (1.42)	10.79 (1.19)	17.07 (2.26)	4.04 (0.23)	4.14 (0.29)	7.99 (0.74)	0.80 (0.20)	11.59 (5.77)	2.87 (0.07)	14.02 (1.23)	9.21 (1.23)
Hexanal	Compound 4	1.66 (0.93)	34.30 (9.74)	25.5 (2.81)	40.73 (28.81)	54.71 (63.10)	12.51 (9.66)	53.37 (0.92)	24.70 (3.28)	21.07 (0.81)	53.73 (7.38)	23.75 (3.59)	27.24 (4.04)	79.50 (5.18)
2-Furanmethanol	Compound 5	14.66 (13.00)	ND	ND	ND	ND	ND	ND	ND	ND	ND	ND	ND	ND
2-Heptanone	Compound 6	1.98 (1.31)	3.37 (0.39)	1.96 (0.31)	4.53 (1.26)	11.12 (1.72)	2.27 (0.06)	3.02 (0.22)	5.86 (0.33)	1.46 (0.39)	2.59 (0.15)	1.01 (0.93)	2.32 (0.20)	2.32 (0.53)
Heptanal	Compound 7	ND	6.32 (1.32)	3.35 (0.41)	15.01 (0.42)	36.10 (1.35)	4.14 (0.25)	10.14 (0.19)	4.42 (0.09)	4.25 (1.13)	11.53 (1.69)	7.12 (1.08)	9.28 (3.50)	19.74 (3.59)
Octanal	Compound 8	ND	5.19 (0.88)	1.42 (1.24)	13.26 (1.89)	26.30 (4.30)	2.33 (0.12)	7.73 (0.42)	3.26 (0.33)	3.90 (0.65)	9.13 (0.10)	6.36 (1.33)	8.03 (0.08)	21.47 (2.07)
3,5-Octadien-2-ol	Compound 9	ND	4.76 (1.07)	3.23 (0.26)	8.47 (1.27)	17.21 (2.08)	1.62 (0.12)	5.13 (0.22)	2.20 (0.66)	1.96 (0.03)	2.41 (0.44)	1.68 (0.32)	1.70 (0.20)	3.52 (0.33)
5-Butyldihydro-2(3H)thio-phenone	Compound 10	1.39 (2.41)	ND	ND	ND	ND	ND	ND	ND	ND	ND	ND	ND	ND
Nonanal	Compound 11	1.25 (1.38)	6.77 (1.34)	3.61 (0.35)	19.25 (3.23)	25.28 (5.09)	7.10 (0.79)	13.91 (1.94)	7.75 (0.86)	11.17 (2.33)	13.46 (0.56)	13.47 (2.98)	11.83 (1.63)	24.75 (2.33)
4H-Pyran-4-one, 2,3-dihydro-3,5-dihydroxy-6-methyl-	Compound 12	6.68 (6.10)	ND	ND	ND	ND	ND	0.54 (0.94)	1.25 (1.17)	0.51 (0.89)	ND	ND	ND	ND
Decanal	Compound 13	ND	1.06 (0.30)	0.40 (0.35)	6.25 (0.81)	5.96 (1.26)	1.56 (0.59)	2.47 (0.82)	0.38 (0.66)	2.69 (1.90)	1.52 (0.08)	3.36 (0.98)	2.34 (0.23)	3.28 (1.01)
2-Furancarboxaldehyde, 5-	Compound 14	4.67 (8.09)	ND	ND	ND	ND	ND	ND	ND	ND	ND	ND	ND	ND

Values in parentheses are standard deviations.

Aldehydes have been previously associated with flavor degradation in SMP and maltol has been established as one of the key volatile compounds contributing to fresh SMP flavor (Karagul-Yuceer et al., 2004; Carunchia-Whetstine and Drake, 2006). Thus, there are some clear observations that can be made about volatile compounds and sensory flavor perception, and some visible similarities in how each analysis differentiated the SMP. The control 1-month SMP is distinct from the other SMP and SMP F3 9 months, F2 30 months, and F4 30 months were distinct from the other SMP by both sensory and instrumental analysis. However, SMP F1 30 months, which was quite distinct by sensory analysis, was not distinct by instrumental analysis. SMP F7 9 months was also differentiated from the other SMP by sensory analysis and not by instrumental analysis. Further, observation of the principal component biplots also suggests that instrumental analysis resulted in more differentiation of the majority of the SMP from the control than sensory analysis. These issues emphasize once again the drawbacks of relying on instrumental results without the clarification of sensory analysis.

IV. Conclusions

Establishing clear relationships between instrumental and sensory analyses requires both instrumental and sensory analysis. The labor-intensive systematic approach remains the optimum way to establish and confirm precise relationships. Less precise but often equally powerful insights can be gained without GC-O, threshold, or model system testing, but both sensory and instrumental analysis are still required.

Acknowledgments

Funding was provided in part by the California Dairy Research Foundation and Dairy Management, Inc. The use of trade names does not imply endorsement or lack of by those not mentioned.

References

ADPI, 2002. *Ingredient Description Brochure*. American Dairy Products Institute. Elmhurst, IL.

ASTM, 1992. Standard practice for determination of odor and taste thresholds by a forced-choice method of limits, E-679-91. In *Annual Book of Standards*, Vol. 15.07. American Society for Testing and Materials, Philadelphia, PA, pp. 35–39.

Avsar, Y.K., Karagul-Yuceer, Y., Drake, M.A., Singh, T., Yoon, Y., and Cadwallader, K.R., 2004. Characterization of nutty flavor in Cheddar cheese. *J. Dairy Sci.* 87: 1999–2010.

Bodyfelt, F.W., Tobias, J., and Trout, G.M., 1988. Sensory defects of dairy products: an overview. In *The Sensory Evaluation of Dairy Products*. Van Nostrand Reinhold: London, pp. 67–89.

Carunchia, M., Parker, J.D., Drake, M.A., and Larick, D.K., 2003. Impact of starter culture rotation on flavor of liquid Cheddar whey. *J. Dairy Sci.* 86(2): 439–448.

Carunchia-Whetstine, M.E., Cadwallader, K.R., and Drake, M.A., 2005. Characterization of rosey/floral flavors in Cheddar cheese. *J. Agric. Food Chem.* 53: 3126–3132.

Carunchia-Whetstine, M.E., Croissant, A.E., and Drake, M.A., 2005. Characterization of WPC80 and WPI flavor. *J. Dairy Sci.* 88: 3826–3829.

Carunchia-Whetstine, M.E. and M.A. Drake., 2006. The flavor and flavor stability of skim and whole milk powders. In *Flavor Chemistry of Dairy Products*. K.R. Cadwallader, M.A. Drake, R. McGorrin, Eds. ACS Publishing, Washington, D.C., chap. X, in press.

Caudle, A.D., Yoon, Y., and Drake, M.A., 2005. Influence of flavor variability in skim milk powder on consumer acceptability of ingredient applications. *J. Food Sci.* 70: S427–S431.

Drake, M.A. and Civille, G.V., 2003. Flavor Lexicons. *Compr. Rev. Food Sci.* 2(1): 33–40.

Drake, M.A., Karagul-Yuceer, Y., Cadwallader, K.R., Civille, G.V., and Tong, P.S. 2003. Determination of the sensory attributes of dried milk powders and dairy ingredients. *J. Sensory Stud.* 18: 199–216.

Drake, M.A., 2004. Defining dairy flavors. *J. Dairy Sci.* 87: 777–784.

Drake, M.A., Cadwallader, K.R., and Carunchia-Whetstine, M.E., 2006. Establishing links between sensory and instrumental analysis of dairy flavors. *Flavor Chemistry of Dairy Products*. K.R. Cadwallader, M.A. Drake, R. McGorrin, Eds. ACS Publishing, Washington, D.C., chap. X, in press.

Drake, M.A., 2006. Defining cheese flavor. In *Improving the Flavour of Cheese*, B. Weimer, Ed., Woodhead Publishing Cambridge, UK.

Engel, W., Bahr, W., and Schieberle, P., 1999. Solvent assisted flavor evaporation — a new and versatile technique for the careful and direct isolation of aroma compounds from complex food matrices. *Eur. Food Res. Technol.* 209, 237–241.

Friedrich, J.E. and Acree, T.E., 1998. Gas chromatography olfactometry of dairy products. *Int. Dairy Journal.* 8(3): 235–231.

Grosch, W., 1993. Detection of potent odorants in foods by aroma extract dilution analysis. *Trends Food Sci. Technol.* 4, 68–73.

Guth, H. and Grosch, W., 1994. Identification of the character impact odorants of stewed beef juice by instrumental analyses and sensory studies. *J. Agric. Food Chem.* 42, 2862–2866.

Karagul-Yuceer, Y., Drake, M.A., and Cadwallader, K.R., 2001. Aroma active components of nonfat dried milk. *J. Agric. Food Chem.* 49: 2948–2953.

Karagul-Yuceer, Y., Cadwallader, K.R., and Drake, M.A., 2002. Volatile flavor components of stored nonfat dry milk. *J. Agric. Food Chem.* 50: 305–312.

Karagul-Yuceer, Y., Drake, M.A., and Cadwallader, K.R., 2003. Aroma active components of liquid Cheddar whey. *J. Food Sci.* 68: 1215–1219.

Karagül-Yüceer, Y., Drake, M.A., and Cadwallader, K.R., 2004. Evaluation of the character impact odorants in skim milk powder by sensory studies on model mixtures. *J. Sensory Stud.* 19: 1–14.

Kirchhoff, E. and Schieberle P., 2001. Determination of key aroma compounds in the crumb of a three-stage sourdough rye bread by stable isotope dilution assays and sensory studies. *J Agric. Food Chem.* 49: 4304–4311.

Lawless, H., Harono, C., and Hernandez, S., 2000. Thresholds and suprathreshold intensity functions for capsaicin in oil and aqueous based carriers. *J. Sensory Stud.* 15: 437–447.

McGorrin, R.J., 2001. Advances in dairy flavor chemistry. In *Food Flavors and Chemistry: Advances of the New Millennium*. Spanier, A.M., Shahidi, F., Parliment,

T.H., Mussinan, C.J., Ho, C.T., and Contis, E.T., Eds. Royal society of Chemistry: Cambridge, pp. 67–84.

Meilgaard, M.C., Civille, G.V., and Carr, B.T., 1999. *Sensory Evaluation Techniques,* 3rd ed., CRC Press: Boca Raton, FL.

Parliament, T.H. and McGorrin, R.J., 2000. Critical flavor compounds in dairy products. In *Flavor Chemistry: Industrial and Academic Research.* Symposium series Nr 756. Risch, S.J. and Ho, C.T., Eds. American Chemical Society: Washington D.C., pp. 44–71.

Peterson, D.G. and Reineccius, G.A. 2003a. Characterization of the volatile compounds that constitute fresh sweet cream butter aroma. *Flavor Fragrance J.* 18: 215–220.

Peterson, D.G. and Reineccius, G.A. 2003b. Determination of the aroma impact compounds in heated sweet cream butter. *Flavor Fragrance J.* 18: 320–324.

Rehman, Shakeel-Ur, Nana Y. Farkye, and Andrew A. Schaffner, 2003. The effect of multiwall Kraft paper or plastic bags on physico-chemical changes in milk powder during storage at high temperature and humidity. *Int. J. Dairy Technol.* V56(1): 12–16.

Rychlik, M., Schieberle, P., and Grosch, W., 1998, Compilation of thresholds, odor qualities, and retention indices of key food odorants. Deutsche Forschungsanstalt fur Lebensmittelchemie and Instiut fur Lebensmittelchemie der Technischen Universitat Munchen. Garching, Germany.

Scheiberle, P. and Hofman T., 1997. Evaluation of the character impact odorants in fresh strawberry juice by quantitative measurements and sensory studies on model mixtures. *J. Agric. Food Chem.* 45: 227–232.

Schieberle, P., Gassenmeier, K., Guth, H., Sen, A., and Grosch, W., 1993. Character impact compounds of different kinds of butter. *Lebensm. Wiss. Technol.* 26: 347–356.

Singh, T., Drake, M.A., and Cadwallader, K.R., 2003. Flavor of Cheddar cheese: a chemical and sensory perspective. *Compr. Rev. Food Sci.* 2: 139–162.

USDEC, 2001. *Reference Manual for U.S. Milk Powders.* U.S. Dairy Export Council, Arlington, VA, pp. 23–29.

Van den Dool, H. and Kratz, P., 1963. A generalization of the retention index system including linear programmed gas liquid partition chromatography. *J. Chromatogr.* 11: 463–471.

Van Ruth, S. 2001. Methods for gas chromatography-olfactometry: a review. *Biomol. Eng.* 17: 121–128.

Varnum, A.H. and Sutherland, J.P., 1994. *Milk and Milk Products: Technology, Chemistry, and Microbiology.* Chapman and Hall, London.

Walstra P., Geurts T.J., Noomen A., Jellema A., van Boekel M.A.J.S., 1999. Milk powder. In *Dairy Technology: Principles of Milk Properties and Processes.* New York: Marcel Dekker, pp. 445–470, chap. 17.

Wright, J.M., Carunchia-Whetstine, M.E., Miracle, R.E., and Drake, M.A., 2006. Characterization of a cabbage off-flavor in whey protein isolate. *J. Food Sci.,* 71: C91–96.

chapter three

Application of sensory-directed flavor-analysis techniques

Ray T. Marsili

Contents

Two common problems confront flavor chemists: (1) determining the cause of an off-flavor in a food or beverage product, and (2) identifying the chemical(s) responsible for a desirable flavor in a natural product or a rival product. The chemicals that provide the principal sensory identity of a product are commonly referred to as character-impact chemicals. A relatively simple mixture of just a few chemicals often defines a food's characteristic flavor.

There are many ways for flavor chemists to study character-impact chemicals. The most appropriate techniques and tools to use depend on the nature of the problem, the type of sample, the analytes of interest, how quickly results are needed, instrumentation availability, the importance of the problem, and other factors.

I. Where to start

First, it is important to collect as much background information about the problem as possible. If the flavor problem is related to elucidation of an off-flavor in a particular food sample, preliminary questions might include the following:

- Is this a common problem that has occurred in the past? Is the problem seasonal? Does it occur with samples manufactured in a specific geographical region or a specific plant?
- Is the product's odor indicative of the taste problem, or is the problem only perceptible when the product is actually tasted? In the former case, gas chromatography-mass spectrometry (GC-MS) studies should be a major part of the analytical strategy to understand the problem; in the latter case, the taste problem may be related to inorganic chemicals (like sodium chloride or other salts) or relatively nonvolatile chemicals (e.g., peptides, free amino acids, sugars, organic acids, etc.), and HPLC or other analytical techniques may be more appropriately applied.
- How do customers describe the off-flavor? How do trained sensory scientists describe the odor or taste? What can be learned from the actual off-flavor descriptors expressed? Do the descriptors indicate the possibility of lipid oxidation, cooked notes from overheating during processing, etc.?
- Has anything unusual happened at the processing plant? For example, have processing temperatures been changed? Has a new ingredient supplier been used? Are new packaging materials involved?
- Has microbiological testing also been conducted? Do results indicate off-flavors are caused by microbial metabolites?
- Is this a single, isolated complaint sample, or is the whole production lot contaminated with the malodor? If only a portion is contaminated, is it at the beginning, middle, or end of the run?
- Is an accelerated shelf life study warranted? Well-planned shelf life studies could, for example, provide information that will help in determining if an off-flavor is the result of incompatibility issues between the packaging and the food product. Shelf life studies can reveal if the loss of desirable flavor chemicals into or through the packaging barrier film has occurred (scalping); if solvents, plasticizers, antistatic chemicals, polymerization accelerators, cross-linking agents, or other chemicals in the packaging materials are being transferred to

the product (leaching); or if lipid oxidation, Maillard browning, or other undesirable chemical reactions are occurring over time.

Even at this early stage of investigation, notice how important sensory input is to the off-flavor resolution process. Most of these questions must be answered with sensory input. Right from the beginning of investigations, the flavor chemist formulates analytical methodologies and strategies based on important sensory information. The higher the quality of this sensory input, the more likely the off-flavor problem will be solved.

II. Sample preparation

After careful consideration of background information, the next step will be to select an appropriate analytical sample-preparation technique. Answers to the questions posed earlier can provide clues as to which sample-preparation method to select. Deciding how to extract, isolate, and concentrate sample analytes is one of the most important steps in the entire off-flavor resolution process; pick the wrong analytical sample-preparation method, and failure is inevitable from the start. Much has been written describing the advantages, disadvantages, and pitfalls of various analytical methodologies and how to select appropriate methods for specific analytes and types of samples [1–11].

Although an oversimplification of the problem, Table 3.1 provides a starting point to assist in the selection of an appropriate sample-preparation technique and lists some of the more common sample-preparation techniques for GC-MS. Because character-impact flavor chemicals are usually volatiles or semivolatiles, sample extraction followed by GC-MS is the most common analytical strategy for flavor analysis. The complexity of the samples being analyzed typically demands that some type of high-resolution chromatographic separation be achieved before the component analytes can be measured and characterized. Capillary GC normally provides sufficient resolving power for analyzing most sample extracts. Even under ideal conditions, it is frequently impossible to resolve all the components in complex food samples. Because of this, analysts often spend a considerable amount of time optimizing parameters and developing methods to achieve effective, meaningful separation of targeted analytes. In some cases, it is necessary to resort to multidimensional GC [12] or GC-time of flight MS (GC-TOFMS) instruments in combination with sophisticated peak-deconvolution software algorithms [13] to achieve the desired resolution of chromatographic peaks.

Every sample-preparation method has a bias [14]. Although most flavor chemists stress that the sample-preparation method should provide an accurate representation of all the volatiles present in the sample's headspace, this criterion may be overrated. It is not only an extremely difficult, if not impossible, goal to accomplish, but it is usually quite unnecessary. The most important criterion for an analytical method is that it is capable of accurately measuring the chemical(s) responsible for flavor. While a food sample may contain hundreds of volatiles, in most cases, only a few (< 10)

Table 3.1 General Comparison of Common Analyte-Extraction Techniques for Studying Food Aromas

Method	Samp. matrix	Samp. size (g)	Detect. Limit	Range of Volatiles				Prep Time for Single Sample
				Gas	Vol.	Semi-Vol.	Non-Vol.	
Static HS[a]	G/L/S	0.1–10	ppm					5–30 min
Dynamic HS/P&T[b]	L/S	1–1000	ppm–ppt					10–30 min
SPME[c]	G/L/S	0.1–100	ppm–ppt					5–60 min
Solvent Extr	L/S	0.1–10	ppb					>30 min
SFE[d]	S	0.1–10	ppb					10–60 min
DTD[e]	S	0.001–0.1	ppb					5–20 min
SBSE[f]	L	0.1–100	ppt–ppq					
HSSE[g]	G/L/S							1–24 hr

Boiling point (°C): 0   100   200   300

[a]Headspace
[b]Purge-and-Trap
[c]Solid-phase microextraction
[d]Supercritical fluid extraction
[e]Direct thermal desorption
[f]Stirbar Sorbent Extraction based on Gerstel Twister® PMDS sorption (Twister immersed in liquid or diluted sample)
[g]Headspace Sorbent Extraction based on Gerstel Twister® PMDS sorption (Twister placed in headspace above sample)
Note: Actual parameters will vary depending on Sample Matrix, Analyte, Type of GC Detector, etc.

G = gas; L = liquid; S = solid; ppq = parts per quadrillion; ppt = parts per trillion; ppb = parts per billion; ppm = parts per million.

character-impact compounds define the sample's typical flavor. Consider, for example, the analysis of pickle brine to identify character-impact chemicals that is discussed later in this chapter. Even though there are dozens of volatiles in fermented pickle brine, only *trans*-4-hexenoic acid and phenyl ethanol are required to mimic the characteristic odor of the brine.

III. Identifying character-impact components

Once you have decided on an initial sample-preparation method, your next challenge is to identify the chemical(s) in your chromatogram responsible for the odor or taste problem. This can be simple, complicated, or nearly impossible. Figure 3.1 illustrates a typical dilemma facing the flavor chemist. Today's flavor chemists have such excellent sample-preparation tools, as well as sophisticated GC-MS systems equipped with autosamplers and powerful chromatographic software, that they can generate dozens of sample chromatograms containing hundreds of chemical peaks in a relatively short time. Figure 3.1 shows a chromatogram of a cheese-flavored cracker containing over 100 volatiles. Other types of samples (e.g., coffee) may have several hundred volatile peaks. How do you determine which ones refer to character-impact chemicals? More data (i.e., more chemical peaks) do not necessarily improve chances of solving flavor problems. Finding character-impact chemicals is like looking for a needle in a haystack. What is often required, of course, is to supplement the analytical process with some type of sensory input.

Comparing chromatograms of control (normal tasting) or "gold-standard" samples with those of complaint samples is a good place to start. Differences in chromatograms can sometimes instantly reveal what has occurred to cause the off-flavor. (See example of complaint and control ice cream samples in Chapter 1, Figure 1.2.)

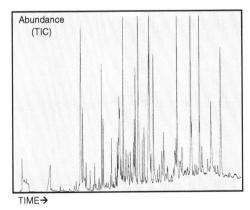

Abundance (TIC)

TIME→

How to...

- *Quantitatively extract/isolate character impact flavor chemicals from sample.*

- *Find peaks important to flavor.*

- *Evaluate (or rank) the relative flavor contribution of each peak.*

- *Understand how each flavor chemical impacts overall flavor perception of food product.*

- *Find meaningful trends in GC-MS data to help resolve flavor issues.*

Figure 3.1 Analytical challenges in flavor analysis: GC-MS chromatogram of a cheese-flavored cracker showing over 100 chemical components.

A.   Olfactometry

A simple comparison of control and complaint samples may fail to determine
the chemical(s) responsible for an off-flavor. Figure 3.2 shows a flowchart to
assist with identification of odor-active chemicals in a complaint sample. An
excellent ancillary technique to help provide sensory insights is olfactometry.
Chapter 4 discusses olfactometry techniques in detail, and many other excellent books and articles describe the techniques [15–22].

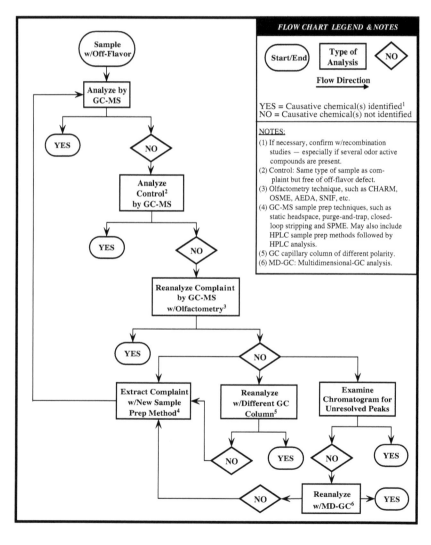

Figure 3.2 Decision tree: Identification of odor-active chemicals in complaint food
sample with off-flavor.

Sniffing GC effluents of control and complaint samples can sometimes reveal the causative agent. When this fails, three options should be considered:

1. Reanalyze samples with the original sample-preparation technique but use a different GC column. Using a column of different polarity may sometimes reveal peaks that were not apparent with the original column selected. (See the application example involving beet sugar malodor at the end of this chapter.)
2. Reexamine chromatograms for the possibility of unresolved peaks. If unresolved peaks are suspected, samples can be reanalyzed using multidimensional GC-MS (see Chapter 3). Another approach that is sometimes used is to separate different classes of chemicals (e.g., into acidic, basic, and neutral fractions) in your sample-preparation technique prior to performing GC-MS. By separating odorants into several fractions prior to GC-olfactometry experiments and analyzing the fractions separately, coelution of odorants from the GC column can often be avoided. This approach has been previously described by T.H. Parliment [23] and is discussed by M. Qian et al. in Chapter 5.
3. Reextract samples using a different sample-preparation technique in case the causative chemicals were not extracted by the first sample-preparation method attempted. Applying different polarity GC columns or multidimensional GC-MS may also be necessary within the extraction technique.

B. Multivariate analysis

Multivariate analysis (MVA) is a powerful technique for understanding how myriad chemicals in a product may impact flavor and odor. MVA provides a way to examine the large amount of peak-area data generated in a typical GC-MS analysis of a food product to distinguish between meaningful information and random variation in the data set. Resolving this problem with traditional univariate methods based on simple algorithms can be difficult, or even impossible, because important relationships exist between combinations of variables (e.g., GC-MS peak areas). In contrast, MVA methods examine many variables simultaneously and attempt to reduce the number of factors (linear combinations of independent variables) that contain the maximum amount of information. In general, the goal of MVA methods is to reduce the dimensionality of the data set, enabling (1) classification of individual samples that make up the data set according to their degree of similarity or (2) prediction of some type of continuous property of samples (e.g., a food product's shelf life or flavor score).

MVA generally, starts with the application of exploratory algorithms to the data set. Patterns of association exist in many data sets, but the relationships between samples can be difficult to discover when the data matrix exceeds three or more features. Exploratory data analysis can reveal hidden patterns in complex data by reducing the information to a more comprehensible format.

This sort of chemometric analysis can expose possible outliers and indicate whether there are patterns or trends in the data. Exploratory algorithms such as principal component analysis (PCA) and hierarchical cluster analysis (HCA) are designed to reduce large complex data sets into a series of optimized and interpretable views. These views emphasize the natural groupings in the data and show which variables most strongly influence those patterns.

Once PCA or HCA show discernible patterns of association between samples, the next step is to apply MVA classification methods or quantitative methods that measure some continuous property of the samples of interest.

Some flavor applications require that samples be assigned to predefined categories or "classes." This may involve determining whether a sample is good or bad (e.g., it has an off-flavor or malodor) or predicting an unknown sample as belonging to one of several distinct groups — e.g., a bakery product with cooked notes, soy oil samples with oxidized flavor, cereal products with beany off-notes, meat samples with warmed-over flavor (WOF), etc. A classification model is used to predict a sample's class by comparing the sample to a previously analyzed experience set where the categories are already known. Two methods — k-nearest neighbor (KNN) and soft-independent modeling of class analogy (SIMCA) — are primary chemometric workhorses used for sample classification. When these techniques are used to create a classification model, the answers provided are more reliable and include the ability to reveal unusual samples in the data. In this manner, an objective chemometric system can be built, thereby standardizing the data evaluation process.

MVA can also be used in flavor chemistry applications to make quantitative predictions on shelf life, flavor score, or some other continuous property of a food product. The goal of chemometric-regression analysis is to develop a calibration model which correlates the information in the set of known measurements to the desired property. Chemometric algorithms for performing regression analysis include partial least squares (PLS) and principal component regression (PCR) and are designed to avoid problems associated with noise and correlations in the data. Because the regression algorithms used are based on factor analysis, the entire group of known measurements is considered simultaneously. Information about correlations among the variables is automatically built into the calibration model.

In the field of food and flavor chemistry, methods of correlating sensory and analytical data represent a very useful application of MVA. By applying MVA procedures, it is often possible to establish meaningful correlations between subjective sensory data and objective instrument data. The ultimate aim is to understand how differences in organoleptic properties among a range of samples are caused by variations in chemical composition.

A minority of foods are characterized by well-defined sensory attributes originating from one or two specific odor-impact compounds. In such cases, correlation of sensory and instrumental data is relatively straightforward. For most products, however, the situation is significantly more complicated because perceived sensory characteristics generally result from several chemical constituents acting in concert. Organoleptic properties are a multivariate phenomenon, so the task of characterizing them realistically requires multivariate methods.

MVA is especially useful when a customer-complaint sample contains flavor chemicals that are normally present, but an off-flavor exists because the ratios of odor-active chemicals are imbalanced. For an example of this type of MVA application, see the example of cheese powders with an off-flavor problem described later in this chapter.

Many types of flavor problems can be resolved with chemometrics and MVA. A few examples include:

1. Classifying samples based on similarity in flavor profiles [24].
2. Classifying samples by type of off-flavor mechanism [25,26].
3. Classifying samples into one of several categories: good vs. bad; control samples vs. complaint samples; Type A vs. Type B vs. Type C; Cabernet Sauvignon wine samples produced at different vineyards; ground coffee from beans of different countries of origin [27].
4. Flavor score prediction.
5. Shelf life prediction (prediction of the number of days after production that the sample will develop unacceptable flavor or taste attributes) [28,29].

Figure 3.3 shows a decision tree that can be used in selecting the appropriate pattern-recognition method(s) for a given problem [30]. For further explanation and examples of how chemometrics and MVA can be used to resolve flavor and off-flavor issues, see Chapters 7 and 8.

C. Verify your hypothesis: recombination studies and odor units

Based on analytical, GC-O, MVA, and sensory studies, you may be able to formulate a theory as to what chemicals are causing an off-flavor and how they were created in the complaint sample. At this point, you should attempt to verify your hypothesis. Unfortunately, in many cases, this critical step in flavor resolution is overlooked or ignored entirely.

One approach is to perform recombination studies. (This is sometimes referred to as "model system" studies.) Application examples demonstrating the use of recombination studies to resolve flavor problems are described later in this chapter. Basically, the process involves adding varying amounts and combinations of potential odor-active compounds identified in your sample to a base or matrix comprised of the major ingredients in the study sample to see if you can match the off-flavor or malodor. The base should duplicate the major ingredients in your off-flavor sample — minus the offending off-flavor chemical(s), of course. You should perform quantitative analysis of your suspect odor-active chemicals to determine the range of concentrations to investigate. You could also spike a control sample (i.e., a sample with normal flavor) of your product with possible suspect candidates to see if you can create the observed off-flavor in the control.

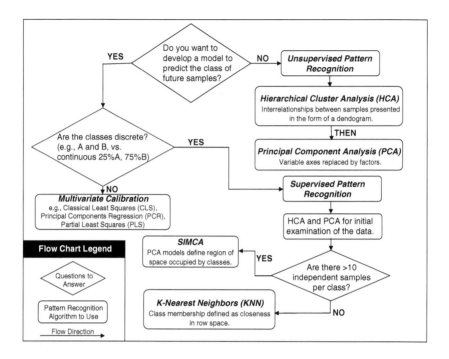

Figure 3.3 Decision tree for choosing the appropriate pattern-recognition (MVA) method(s). (From K.R. Beebe, R.J. Pell, and M.B. Seaholtz, *Chemometrics: A Practical Guide*, John Wiley & Sons, New York, 1998, p. 63. With permission.)

Another technique that is sometimes used to verify that the chemicals you have identified are, indeed, the causative agents is to calculate odor units (also known as odor-activity value or OAV, unit flavor base, odor value, or aroma value) [31–36]. The odor unit (U_o) of a chemical is defined as follows:

U_o = volatile concentration in sample/odor threshold of the volatile in water.

Sometimes, taste units are calculated. The taste unit (U_t) of a chemical is defined as follows:

U_t = volatile concentration in sample/taste threshold of the volatile in water.

Some researchers prefer to calculate the logarithm of the odor unit (log U_o) to represent changes in concentration which are significant for olfactometry discrimination. Odor activity follows a sigmoidal dose-response curve because significant aroma responses require order-of-magnitude changes in concentration. Consequently, logarithmic functions more significantly represent meaningful sensory differences. Log U_o values > 1 (which signify that the volatile is present at levels > 10 times higher than its odor threshold level) are indicative of compounds present at concentrations that greatly exceed their thresholds and, therefore, are likely to contribute significant flavor impact [34].

Sensory threshold values are highly dependent on the sample matrix, pH, sample temperature, and experimental techniques used to determine threshold levels. Although many flavor chemists use odor units based on odor-threshold values determined in water, this may be inappropriate for many types of analytes and matrices. Ideally, sensory-threshold data should be determined in the same matrix as the sample being studied.

For example, if odor units are being calculated for volatiles in a cheese sample, odor thresholds for the various analytes should be determined in a cheese matrix. One obstacle in using odor units to understand the degree of flavor contribution of various chemicals in a food product is that odor-threshold values for various organic chemicals are not published in many types of matrices. Odor-threshold values for chemicals in cheese can be determined experimentally, but this is a very time-consuming task. Even though it may be an oversimplification, basing U_0 calculations on odor-threshold values determined in water is common practice, because tables of odor-threshold values for numerous organic chemicals have been published in the literature and are readily available.

When the analytes in the sample tend to partition strongly in an oil phase, some flavor chemists attempt to use odor-threshold values determined in vegetable oil. Some published data for thresholds of various organic chemicals in vegetable oil are available [33].

Calculating odor units can assist in the process of identifying the most potent odorants and eliminate odorants that are probably not involved in the off-flavor problem. For example, with the cheese-powder flavor problem discussed later in this chapter, odor units were calculated for butyric, isovaleric, hexanoic, heptanoic, octanoic and decanoic acids. All of these organic acids had high log U_0 values, indicating that they were important contributors to the flavor of this product.

Odor units have also been used to prioritize odor intensity of chemicals when it was not practical or advisable to perform olfactometry experiments. R.J. McGorrin and L. Gimelfarb attempted to isolate and identify the potent volatile and nonvolatile flavor components in fresh and cooked tomatillos and to compare similarities and differences to those previously reported for tomatoes [34].

An informal sensory panel comprising of four experienced members trained in the Sensory Spectrum method was used to evaluate flavor-profile differences among fresh and cooked samples. Descriptor terms were generated by the panelists and intensity scores were anchored on a four-point scale from least to most. Dynamic headspace GC-MS was used to identify volatile components in samples.

As previously discussed, GC-O techniques such as aroma-extraction dilution analysis (AEDA) [15] or Charm analysis [37] are often applied to elucidate the key odorants in flavor isolates. However, GC-O techniques are somewhat time-consuming and require repetitive analyses. In this case, the researchers were concerned that relative concentrations of flavorants could change during the time interval required for multiple analyses and decided to use U_o measurements to identify the most important odorants in tomatillo. Using published odor-threshold values of organic chemicals in aqueous solutions, the researchers were able to calculate U_o values for most of the volatiles in tomatillo samples. Many of the volatiles identified in the tomatillo study have been previously reported in odor-threshold tables. In fresh tomatillo, (Z)-3-hexenal had the highest log U_o value (4.1), followed by nonanal (2.8), hexanal (2.3), β-damascenone (3.2), and decanal (2.5).

M. Qian and G. Reineccius identified and quantitated potentially important aroma compounds in Parmigiano-Reggiano cheese and then used calculated OAVs (odor units) based on sensory thresholds reported in literature to identify the chemicals most important to the flavor of Parmigiano-Reggiano cheese [36].

OAVs were calculated from published threshold data for volatiles in various media, including water, vegetable oil, butter, milk, cheese, etc. The researchers found that 3-methylbutanal, 2-methylbutanal, 2-methylpropanal, dimethyl trisulfide, diacetyl, methional, and phenylacetaldehyde were important contributors to the aroma of Parmigiano-Reggiano cheese. Other important aroma compounds included several ethyl esters of free fatty acids (C4:0, C6:0, C8:0), as well as acetic acid and several free fatty acids (C4:0, C6:0, C8:0).

One drawback to performing both recombination studies and odor-unit experiments is that levels of odor-active chemicals must be accurately quantitated in the sample of interest. The quantitation process can be time-consuming and difficult to perform for some types of sample-preparation methods. However, when time is available and the problem being studied is important enough, the rewards are worth the effort.

IV.   Application examples

The following case histories illustrate how olfactometry, multivariate analysis, recombination (model system) studies, and consideration of OAVs have been used to help resolve different kinds of flavor problems in a variety of food systems.

A.   Fermented pickle flavor

Most dill pickles are produced by fermenting cucumbers in large vats containing brine solution. Reducing sugars are fermented to lactic acid and other secondary metabolites, some of which impart the characteristic flavor to fermented dill pickles. After a few weeks, fermentation is complete, and the pickles are removed from the brine and packed in jars or other containers with dill oil, garlic oil, and other spices. The fermentation process is a relatively expensive part of processing because it ties up inventory for several months and generates a considerable amount of brine waste that can be taxing to local sanitary districts.

To eliminate these expenses, a pickle processor developed a way to make dill pickles without the costly fermentation step [38]. The nonfermented pickles had a fresher, green flavor compared to traditional fermented pickles. While most people preferred the fresher flavor and crunchier texture of the nonfermented pickles, the major customer of the product, a large international hamburger retailer, rejected the new product because it tasted different from the pickles its customers had become accustomed to. The company's flavor chemists were told to find out what chemicals were responsible for the characteristic fermented dill-pickle flavor so that these flavor chemicals could be added to the nonfermented product.

Although university researchers had been attempting to identify the odor-impact chemicals of fermented pickles for many years, no one could solve the problem. (The major reason why they were unable to identify character-impact chemicals in pickle brine was that they selected a sample-preparation or isolation technique [purge-and-trap on Tenax] that failed to capture the major character-impact chemicals in brine.) Over a period of several months, the flavor chemists collected numerous fermented pickle samples from four processing plants located in different states. All brine samples collected had the characteristic fermentation odor. However, the intensity of the odor varied from location to location.

Initial GC-MS analysis using purge-and-trap techniques with a Tenax-GC trap appeared promising, generating approximately 100 chromatographic peaks. When olfactometry experiments were applied to the purge-and-trap extracts, several odiferous peaks were detected, but none were similar to the pickle brine. Because it was suspected that the key flavor-impact chemicals were not extracted by purge-and-trap, two other sample-preparation techniques were investigated: solid-phase extraction (SPE) with C_{18} cartridges and solid-phase microextraction (SPME) with two types of fibers (75-μm Carboxen/PDMS and 70 μm Carbowax™/DVB Stable Flex™).

Far fewer peaks were observed with SPE, and only a few early-eluting peaks were present in SPME extracts compared to purge-and-trap. Both SPE and SPME showed more late-eluting peaks that were missing from the purge-and-trap results. Preliminary olfactometry experiments showed that several of these late eluters had potent, more interesting odor characteristics that could be responsible for the characteristic fermentation odor. A GC-MS chromatogram of a typical brine solution is shown in Figure 3.4.

A problem with the SPE technique was that degradation of nonvolatile brine components (e.g., chlorophyll and other plant pigments), which were coextracted with the volatile flavor chemicals, tended to elute as broad peaks at the end of chromatographic runs. Therefore, SPE was abandoned and future flavor extracting of brines was accomplished with SPME.

The next step was to employ GC-O. A panel of food technologists was trained to recognize the characteristic fermented brine odor. This was accomplished simply by giving each panelist several fermented pickle brine samples and asking them to compare the odor to a base brine sample consisting of 6,500 ppm lactic acid, 500 ppm acetic acid, and 8% sodium chloride in distilled water.

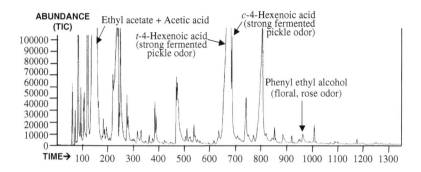

Figure 3.4 SPME GC-MS chromatogram of a typical pickle fermentation brine. (From R.T. Marsili and N. Miller, *J. Chromatogr. Sci.* 38: 307, 2000. With permission.)

The GC-O method selected for this study was the detection-frequency method; details of this method have been previously reported in the literature [18,39,40]. In this technique, the intensity and duration of aromas emitted from the sniff port were recorded for each observed odorant by pressing a button on a switch connected to an A/D converter. The resulting square signal was registered and recorded. Odor assessors verbally described detected odors, and the verbal descriptors and corresponding peak retention times were recorded manually by an assistant. The process resulted in an individual aromagram. The area under each odor peak was obtained by integration software. Aromagrams for the same brine sample were generated by eight different panelists. The eight individual aromagrams were combined to create one master aromagram for the sample. The combined areas for each peak are referred to as Surface Nasal Impact Frequency (SNIF) value [18]. The master aromagram created with the SNIF GC-O method is shown in Figure 3.5.

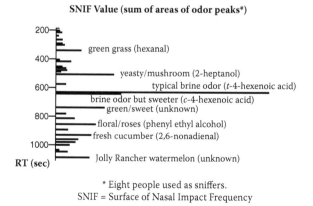

Figure 3.5 GCO-SNIF results of a typical pickle fermentation brine based on combining aromagram SNIF values for eight panelists. (From R.T. Marsili and N. Miller, *J. Chromatogr. Sci.* 38: 307, 2000. With permission.)

The seven odorants with the largest SNIF values were (from highest to lowest) *trans*-4-hexenoic acid, *cis*-4-hexenoic acid, 2-heptanol, *cis*-2,4-hexandienoic acid, phenylethyl alcohol, 2,6-nonadienal, and dodecen-1-al. Based on SNIF values, *trans*-4-hexenoic acid was by far the most powerful odorant observed in brine samples. The *trans*-4-hexenoic acid and *cis*-4-hexenoic acid were detected by all panelists and were the only odorants characterized as "definitely similar to the aroma of the brine sample."

Recombination studies were conducted to determine which of the seven odorants with the largest SNIF values were essential in creating the brine odor. When GC-MS chromatogram data were examined for all samples from different geographic areas, only three chemicals were consistently present in all brine samples with the characteristic brine odor: *trans*-4-hexenoic acid, *cis*-4-hexenoic acid, and phenylethyl alcohol. Because a pure sample of *cis*-4-hexenoic acid was not available, it was not considered in recombination studies. Therefore, recombination studies were based on only two chemicals: *trans*-4-hexenoic acid and phenylethyl alcohol.

Quantitation of eight brine samples (two from four different pickle-processing plants) showed that *trans*-4-hexenoic acid was present in the concentration range of 5 to 114 ppm, and phenylethyl alcohol was present in the 2 to 30 ppm range. Phenylethyl alcohol and *trans*-4-hexenoic acid at various concentrations in this range were added to a base solvent consisting of 6,500 ppm lactic acid, 500 ppm acetic acid, and 8% sodium chloride in distilled water; the chemicals used to make the base solvent were present in all brines tested. Table 3.2, depicting the results of the recombination study, shows that a mixture of 25 ppm *trans*-4-hexenoic acid and 10 ppm phenylethyl alcohol in the base solvent can be used to effectively mimic fermented pickle brine flavor.

This application demonstrates how important selection of the appropriate sample-extraction method is to flavor-matching success, why it is often necessary to investigate more than one sample-preparation method, and why and how olfactometry and recombination experiments can be applied to assist in problem solving.

Table 3.2 Recombination Study Using *t*-4-Hexenoic Acid and Phenylethyl Alcohol Added to a Base Solvent to Create Typical Pickle Brine Odor

Sample	*t*-4-Hexenoic acid (ppm in base[a])	PEA (ppm in base)	Odor match score[b]
A	0	0	0.0
B	2	0	4.0
C	10	0	6.3
D	25	0	7.0
E	25	0.5	7.0
F	25	10	7.7
G	25	40	5.0

[a] Base: 6500 ppm lactic, 500 ppm acetic, and 8% NaCl in DI water.

[b] Three panelists. Match "0" = no match to typical odor; "10" = perfect match.

B. Malodor in beet sugar

A significant quality problem with white beet sugar is a characteristic earthy, musty, barny, silage-like odor often perceived in samples [41]. Although the characteristic aroma defect is not detectable once the sugar is solubilized, it can be extremely strong and objectionable when sealed canisters of the granulated sugar are initially opened by consumers.

Following numerous complaints from its retail customers, a company that marketed beet sugar in canisters wanted to identify the causative agent(s) responsible for the malodor. If the character-impact compounds could be identified, researchers hoped to understand the mechanism of formation of the compounds and devise a way to prevent it from occurring.

The strategy for determining the cause of the characteristic odor defect in beet sugar was based on analytical testing and sensory analysis. First, malodorous sugar samples were analyzed by GC-MS using closed-loop-stripping (CLS) and direct thermal desorption (DTD) sample-preparation techniques. Column effluent was split between the MS detector and an olfactometry detector. Olfactometry experiments with CLS and DTD generated the aromagrams shown in Figure 3.6, Figure 3.7, and Figure 3.8. Beet sugar aromagrams based on olfactometry sniffing experiments were important tools in identifying chemicals that were potential contributors to the characteristic beet odor defect. Figure 3.7 shows a magnified view of the chromatogram in Figure 3.6 in the retention time range of 1600 to 1800 sec. A small but discernible geosmin peak is indicated by MS detection; a 25 ppt spike of odorless cane sugar produced a slightly larger geosmin peak at a retention time corresponding to the detection of the musty odor in the beet sugar samples.

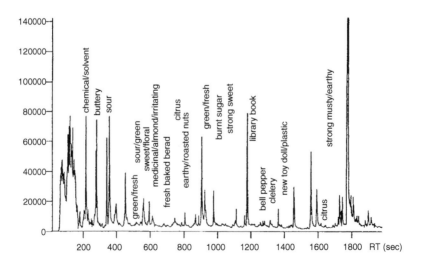

Figure 3.6 Aromagram of beet sugar with typical malodor by GC-O using closed-loop stripping on Tenax TA and MS (ITD) detection. (From R.T. Marsili, N. Miller, G.J. Kilmer, and R.E. Simmons, *J. Chromatogr. Sci.* 32: 165, 1994. With permission.)

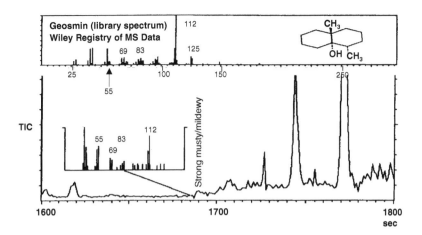

Figure 3.7 Aromagram of beet sugar with typical malodor by GC-O using closed-loop stripping on Tenax TA and MS (ITD) detection in the retention time range of 1600 to 1800 sec. Comparison of library spectrum of geosmin with MS of baseline at the retention time where a strong musty odor was detected in all seven malodorous samples shows trace presence of geosmin. (From R.T. Marsili, N. Miller, G.J. Kilmer, and R.E. Simmons, *J. Chromatogr. Sci.* 32: 165, 1994. With permission.)

Once suspect chemical contributors to the odor problem were identified using the sniff port, the next step was to quantitate levels of the chemicals in several samples of beet sugar with the aroma defect so odor-activity values and recombination studies could be conducted. OAVs and recombination studies provided positive confirmation of the chemicals responsible for the malodor.

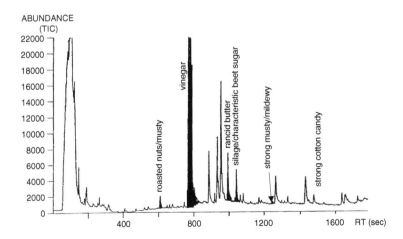

Figure 3.8 Aromagram of beet sugar with typical malodor by GC-O using direct thermal desorption GC-O and MS detection. (From R.T. Marsili, N. Miller, G.J. Kilmer, and R.E. Simmons, *J. Chromatogr. Sci.* 32: 165, 1994. With permission.)

Table 3.3 Concentrations of Potential Odor-Causing Chemicals and Sensory-Panel Scores for Seven Samples of Beet Sugar with Varying Degrees of the Characteristic Musty or Silage Odor Defect

Beet sugar sample	Concentration (ppb)					Sensory odor score[b]
	Geosmin[a]	Furfural	2,5-Dimethyl-pyrazine	Butyric acid	Isovaleric acid	
1	< 0.025	< 10	2.3 ± 0.5[c]	43.4 ± 0.7	19.7 ± 0.9	4.63 ± 0.75
2	< 0.025	< 10	4.9 ± 1.4	63.6 ± 4.0	78.7 ± 4.8	3.88 ± 0.63
3	< 0.025	< 10	15.4 ± 1.9	197 ± 8.0	105 ± 6.0	7.50 ± 0.41
4	< 0.025	< 10	3.4 ± 1.3	247 ± 15	60.2 ± 3.1	8.63 ± 0.48
5	< 0.025	< 10	4.9 ± 1.1	47.2 ± 2.6	61.0 ± 0.5	6.88 ± 1.18
6	< 0.025	< 10	1.7 ± 0.5	990 ± 20	88.2 ± 3.2	7.25 ± 0.5
7	< 0.025	< 10	2.1 ± 0.4	55.9 ± 2.2	32.0 ± 1.7	1.50 ± 0.5

[a] Odors detected with an olfactometry detector indicate all samples contain geosmin, even though the analytical test is incapable of detecting measurable peak areas.
[b] Average sensory odor score, where 0 is no detectable odor and 10 is extremely strong characteristic odor defect; uncertainties represent the standard deviation of scores for four different panelists.
[c] Uncertainties are in ppb units and represent the average deviation of duplicate determinations.
Source: R.T. Marsili, N. Miller, G.J. Kilmer, and R.E. Simmons, *J. Chromatogr. Sci.* 32: 165, 1994. With permission.

Recombination studies were conducted by spiking odorless cane sugar with the suspect chemicals at concentrations similar to those found in the beet sugar. Using this approach, the primary odor-active chemicals in malodorous beet sugar were identified as geosmin (*trans*-1,10-dimethyl-*trans*-9-decalol), furfural, 2,5-dimethylpyrazine, butyric acid, and isovaleric acid.

Concentration levels of major beet-sugar odorants and sensory odor scores (where 0 is no detectable malodor and 10 is extremely strong characteristic odor defect) are shown in Table 3.3, odor thresholds for the major odorants in beet sugar are shown in Table 3.4, and OAVs for these odorants are shown in Table 3.5. Based on OAV data, only one of the suspect odorants in one sample (butyric acid in Sample No. 6) would be expected to contribute odor in the sugar samples. These preliminary findings suggested that the chemicals responsible for the typical malodor were not extracted by the sample-preparation technique employed. To check this hypothesis, recombination studies were conducted.

Table 3.4 Approximate Experimentally Determined Odor Thresholds for Major Odiferous Chemicals Found in Beet Sugar

Odorant	Odor threshold[a] (ppb)	Odor descriptors
Geosmin	0.025	Mild musty/mildew
Furfural	1000	Medicinal, irritating, almondlike
2,5-Dimethylpyrazine	200	Nutty, earthy
Butyric acid	600	Rancid butter, cheesy
Isovaleric acid	400	Pungent, cheesy, silagelike

[a] Odor thresholds determined experimentally by spiking odorless cane sugar samples with varying amounts of individual odorants. Odor thresholds were determined as the lowest concentrations at which three out of five sensory panelists could detect and similarly describe odors.
Source: R.T. Marsili, N. Miller, G.J. Kilmer, and R.E. Simmons, *J. Chromatogr. Sci.* 32: 165, 1994. With permission.

Table 3.5 Odor-Activity Values of Seven Beet-Sugar Samples with Odor Defect

Beet sugar sample	Geosmin	Furfural	2,5-Dimethyl-pyrazine	Butyric acid	Isovaleric acid
1	< 1.0	< 0.01	0.01	0.07	0.05
2	< 1.0	< 0.01	0.02	0.11	0.20
3	< 1.0	< 0.01	0.08	0.33	0.26
4	< 1.0	< 0.01	0.02	0.41	0.15
5	< 1.0	< 0.01	0.02	0.08	0.15
6	< 1.0	< 0.01	0.01	1.70	0.22
7	< 1.0	< 0.01	0.01	0.09	0.08

Source: R.T. Marsili, N. Miller, G.J. Kilmer, and R.E. Simmons, *J. Chromatogr. Sci.* 32: 165, 1994. With permission.

Recombination studies revealed important insights. When sugar samples were spiked with various combinations of odorants, a significant observation was made: mixtures of volatile acids in combination with geosmin produced an aroma identical to the typical odor defect of beet sugar. Furthermore, a synergistic effect was observed — the concentration of acids and geosmin that would produce discernible odor was considerably less than the concentration required when each chemical was added separately to the sugar. In other words, odor thresholds for geosmin and the organic acids were significantly lowered when the geosmin was present in combination with acetic, butyric, or isovaleric acid.

Mixing each acid separately with geosmin produced the characteristic odor defect, irrespective of the specific volatile acid used. However, slight nuances in aromas were observed depending on the type of acid and its concentration. For example, an acetic acid and geosmin combination provided earthy notes, while an isovaleric and geosmin combination had an odor reminiscent of musty silage. A mixture of 25 ppt geosmin, 500 ppb acetic acid, 60 ppb isovaleric acid, and 200 ppb butyric acid provided an excellent match to the typical beet-sugar odor defect. It is interesting to note that the concentration level of each acid in this mixture is well below the odor threshold level observed when each acid (and no geosmin) is added individually to the cane sugar.

Undoubtedly, varying levels of acetic, butyric, and isovaleric acid in combination with geosmin are responsible for slight variations in sample-to-sample aroma characteristics of beet-sugar samples.

When 2,5-dimethylpyrazine or furfural was added to cane sugar spiked with 25 ppt geosmin at concentrations slightly exceeding those found in beet sugar (i.e., 20 ppb 2,5-dimethylpyrazine and 10 ppb furfural), no characteristic beet-sugar odor was observed by any of the sensory panelists. This was true even when concentrations of 2,5-dimethylpyrazine and furfural were doubled to 40 ppb and 20 ppb, respectively.

The chemicals responsible for the characteristic odor defect of beet sugar originate from the action of soil microorganisms on sucrose in sugar beets. Geosmin is produced by numerous actinomycetes and other types of soil molds. Butyric and isovaleric acids are metabolites of soil bacteria (e.g., *Clostridium butyricum* and other species of clostridia). Fructose, glucose, and organic acids have been reported to increase in concentration when beets are stored for prolonged periods prior to processing.

Based on the preceding study, it was possible to develop a strategy to help minimize the intensity of the malodor in sugar beets. Reducing storage time (the interval of time between when beets are harvested and when they are processed) is one way to reduce levels of malodorous metabolites in processed beet sugar. Aeration of processed sugar as a conditioning step prior to shipment has been observed to reduce malodors. In view of the extremely low odor-detection thresholds of the chemicals responsible for beet sugar malodors and the fact that sugar beet is a root crop intimately exposed to soil microorganisms, it may be impossible to completely eliminate the odor defect from commercial beet sugar.

C. Application of MVA in resolving a flavor problem with cheese powder

An ingredient processor was experiencing several rejections of production lots of cheese powder by its customer [42]. Criticisms were inconsistent but frequently implicated cheese flavor problems. The impact of odor-active compounds on cheese flavor is not very well understood. The flavor of cheddar, the most popular cheese flavor in North America, is described as sweet, buttery, aromatic, and walnut, yet there is no general consensus among flavor chemists about the identity of individual compounds or groups of compounds responsible for cheddar flavor. One widely accepted theory, the component balance theory, purports that desirable cheddar flavor is based on a balance of free fatty acids, free amino acids, amino acid catabolites (which include volatile sulfur compounds), and other chemicals.

As a first step to understanding the reason for rejection of cheese powders, the flavor chemist working on the problem analyzed several accepted and rejected cheese powder samples. Since this research involved testing dozens of samples over a prolonged period, automated solid-phase microextraction (SPME) methods were investigated as a sample-preparation technique. A polydimethylsiloxane/divinylbenzene (PDMS/DVB) fiber was selected because it provided good detection sensitivity for a broad range of flavor-impact compounds, including free fatty acids. In addition, several samples were analyzed with a Carboxen/PDMS fiber, which is better suited to analyzing highly volatile compounds such as dimethyl sulfide and acetaldehyde.

Comparison of chromatographic peak areas for numerous accepted and rejected samples did not show much noticeable qualitative or quantitative difference. For example, rejected samples did not contain more or higher levels of chemicals associated with cooked notes, Maillard reaction products, oxidation off-flavors, etc.

Additional testing was conducted using HPLC and other techniques to determine differences in nonvolatile flavor chemicals like sodium chloride, free amino acids, and lactic acid. No significant differences between accepted and rejected samples were observed.

Alternate sample-preparation methods for GC-MS were considered. For example, one method that works particularly well for cheese products is solvent-assisted flavor evaporation (SAFE), which involves a compact distillation unit in combination with a high-vacuum pump and liquid-nitrogen cold trapping [43,44]. After careful consideration, researchers decided not to investigate the SAFE method. Three factors hindered its application for this particular problem: (1) the method was too labor intensive for the number of samples being analyzed (only two or three samples per day could be analyzed with SAFE), (2) the sophisticated glassware setup required for SAFE was not available in the lab and had to be custom-made, and (3) results were urgently needed by the ingredient manufacturer.

Researchers decided to examine the large volume of SPME results more closely using MVA. MVA is an excellent tool for identifying trends in data that may be missed by simple visual inspection or comparison of chromatographic-peak-area results. MVA seemed like a particularly good investigative tool for this problem. If a balance of key odor and flavor chemicals determined good cheddar flavor, then examination of the relative concentrations of these components in accepted and rejected samples could reveal the cause for rejection.

Thirteen samples that passed the customer's sensory screening tests, nine samples that failed the customer's sensory screening tests, and a gold standard (GS) sample (i.e., a sample with an ideal flavor profile as judged by sensory analysis) were collected and analyzed. When two-dimensional principal component analysis (PCA) scores plots were constructed using peak-area data for 60 SPME GC-MS volatiles, clustering of passed (accepted) and failed (rejected) samples was observed, indicating that there was sufficient information contained in SPME results to account for significant differences in the chemical composition of passed and failed samples. Further examination of results showed that free fatty acids (FFAs) had the highest modeling power of all SPME volatiles and were the primary drivers for clustering of passed vs. failed samples. To test the discrimination ability of just the free fatty acid results, a second PCA plot was constructed using only free fatty acid data (Figure 3.9). When FFA concentrations of the 23 samples were plotted in a two-dimensional PCA format, clustering of "P" samples and the "GS" sample was observed, with all "F" samples falling outside this cluster. Therefore, it was possible to classify samples as to their pass or fail status simply by examining PCA plots of FFA data.

It is possible that FFAs were not the primary volatiles responsible for acceptance or rejection of the product by sensory panelists. FFAs may simply have demonstrated cocorrelation with the "true" determinants of product flavor. To determine if it was reasonable that FFAs were major flavor drivers, OAVs and TAVs were determined for most of the acids involved. Typical average levels of FFAs in cheese powder samples (accepted and rejected) are shown in Table 3.6. OAVs and TAVs were calculated by dividing the concentration of the acid in cheese powder by the appropriate threshold value of the acid.

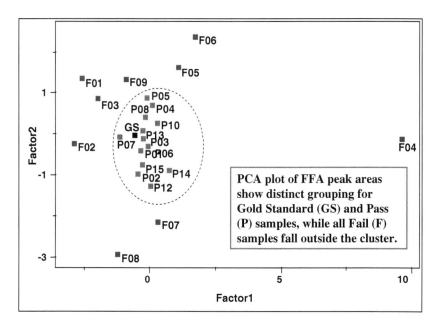

Figure 3.9 PCA plot of SPME GC-MS free fatty acid peak-area data for 13 samples of cheese powder that passed sensory testing, 9 samples that failed, and a GS (ideal flavor profile) sample.

Log OAV and log TAV results showed that the concentrations of FFAs in cheese powders are 10 to more than 1000 times their odor and taste thresholds, making them significant contributors to the odor and taste of cheese powders.

To further test the discrimination capability of FFA levels to determine pass or fail classification of samples, a KNN classification model was developed from the sample set. The KNN model was then used to predict the pass or fail status of 20 new samples — 10 that had passed sensory-analysis

Table 3.6 OAVs and TAVs for Free Fatty Acids in Cheese Powder

| Acid | Concentration in cheese powder (ppm) | Threshold in H$_2$O (ppm) | | Log U$_o$[a] | Log U$_t$[b] |
		Odor	Taste		
C4:0	700	0.24	6.5	3.5	2.0
Iso-C5:0	50	0.12	—	2.6	—
C6:0	500	3.0	5.4	2.2	2.0
C7:0	30	3.0	—	1.0	—
C8:0	225	3.0	5.3	1.9	1.6
C10:0	300	10.0	3.5	1.5	1.9

[a] U$_o$ = volatile concentration/odor threshold in water.
[b] U$_t$ = volatile concentration/taste threshold in water.

screening by the customer and 10 that had failed. The 20 samples were analyzed by SPME GC-MS and the FFA concentrations were determined. The KNN classification model predicted 8 of 10 failed samples correctly and 9 of 10 passed samples correctly. The total percentage of correctly predicted classifications for this data set was 85%.

As a result of this research, the ingredient manufacturer had a relatively reliable and rapid test for checking acceptability of freshly manufactured cheese powder samples prior to shipment to customers. With the new test protocol, the ingredient manufacturer could screen incoming cheeses used in cheese powder formulations to ensure they contained appropriate levels and ratios of FFAs. This example illustrates the potential importance of MVA in the resolution of flavor problems.

V.   Learning from failures

Despite hard work and creative thinking, you may fail to identify the chemicals responsible for a particular off-flavor. You may be able to identify numerous potent odor-active chemicals in your sample but fail to identify any of the candidates as likely contributors to the specific off-odor in your product.

Another particularly frustrating problem is that you may be able to identify the causative agents but fail to determine how they were generated in the complaint sample. Although discovering the chemical responsible for the problem may be scientifically interesting, your real goal is often to understand the mechanism of formation of the offending chemicals and formulate a viable strategy to prevent or minimize their development in future production runs. You may spend more time researching the mechanism of formation of the causative agent than analyzing samples.

Research in determination of characteristic odor or flavor chemicals can be rewarding and frustrating. You may not always resolve the problem. Even when you cannot, your goal should be to learn as much as possible about the flavor components of your product; perhaps the next time a problem arises, your previous experience will help you resolve it. Success in off-flavor resolution is profoundly influenced by researcher experience — knowing what odor-active chemicals are normally present in your sample, how they usually vary from lot to lot, and what off-flavor chemicals are likely to be produced in the product as a result of processing, ingredients, shelf life, and packaging problems. Patience, persistence, experience, and sometimes just plain luck are key factors in successful flavor analysis.

References

1. R. Marsili, Ed., *Techniques for Analyzing Food Aroma*, Marcel Dekker, New York, 1997.
2. R. Marsili, Ed., *Flavor, Fragrance and Odor Analysis*, Marcel Dekker, New York, 2002.
3. B. Kolb, D. Boege, and L.S. Ettre, Advances in headspace gas chromatography: instrumentation and application, *Am. Lab.* 20: 33, 1988.

4. J.W. Washall and T.P. Wampler, Sources of error in purge and trap analysis of volatile organic compounds, *Am. Lab.* 22: 38, 1990.
5. G. Blanch, G. Reglera, M. Herriaz, and J. Tabera, Comparison of different extraction methods for the volatile components of grapefruit, *J. Chromatogr. Sci.* 29: 11, 1991.
6. R. Teranishi and S. Kint, Sample preparation, *Flavor Science: Sensible Principles and Techniques*, T.E. Agree and R. Teranishi, Eds., American Chemical Society, Washington, D.C., 1993, p. 251.
7. J.J. Manura, Direct thermal analysis using the short path thermal desorption system, Short Path Thermal Desorption Note (Scientific Instrument Services, Inc.), No. 5, 1991.
8. A. Krumbein and D. Ulrich, Comparision of three sample preparation techniques for the determination of fresh tomato aroma volatiles, *Flavor Science: Recent Developments*, A.J. Taylor and D.S. Mottram, Eds., The Royal Society of Chemistry Information Services, London, 1996, p. 289.
9. S.S. Wercinski, Ed., *Solid-Phase Microextraction: A Practical Guide*, Marcel Dekker, New York, 1999.
10. X. Yang and T. Peppard, Solid-phase microextraction for flavor analysis, *J. Agric. Food Chem.* 42: 1925, 1994.
11. E. Baltussen, P. Sandra, F. David, and C.A. Cramers, Stir Bar Sorptive Extraction (SBSE), a novel extraction technique for aqueous samples: theory and principles, *J. Microcolumn Sep.* 11: 737, 1999.
12. D.W. Wright, Application of multidimensional gas chromatography techniques to aroma analysis, *Analyzing Food Aroma*, R. Marsili, Ed., Marcel Dekker, New York, 1997, p. 113.
13. J.F. Holland and B.D. Gardner, The advantages of GC-TOFMS for flavor and fragrance analysis, *Flavor, Fragrance and Odor Analysis*, R. Marsili, Ed., Marcel Dekker, New York, 2002, p. 107.
14. G. Reineccius, Biases in analytical flavor profiles introduced by isolation method, *Flavor Measurement*, C.-T. Ho and C. Manley, Eds., Marcel Dekker, New York, 1993, p. 61.
15. W. Grosch, Detection of potent odorants in foods by aroma extract dilution analysis, *Trends Food Sci. Technol.* 4: 68, 1993.
16. R. Rouseff, P. Jella, and J. Lin, Determination of seasonal and thermal processing changes in grapefruit juice aroma active components using GC-olfactometry, *Food Flavors and Chemistry: Advances of the New Millennium*, A.H. Spanier, F. Shahidi, T.H. Parliment, C. Mussinan, C.-T. Ho, and E.T. Contis, The Royal Society of Chemistry, Cambridge, 2001, p. 336.
17. I. Blank, Gas chromatography-olfactometry in food aroma analysis, *Flavor, Fragrance and Odor Analysis*, R. Marsili, Ed., Marcel Dekker, New York, 2002, p. 297.
18. A. Chaintreau, Quantitative use of gas chromatography-olfactometry: The GC-"SNIF" method, *Flavor, Fragrance and Odor Analysis*, R. Marsili, Ed., Marcel Dekker, New York, 2002, p. 333.
19. P. Semmelroch and W. Grosch, Analysis of roasted coffee powder and brews by gas chromatography-olfactometry of headspace samples, *Lebensm. Wiss. Technol.* 28: 310, 1995.
20. J.P.H. Linssen, J.L.G. Janssens, J.P. Roozen, and M.A. Posthumus, Combined gas chromatography and sniffing port analysis of volatile compounds of mineral water packed in polyethylene laminated packages, *Food Chem.* 46: 367, 1993.

21. W. Grosch, U.C. Konopka, and H. Guth, Characterization of off-flavors by aroma extraction dilution analysis, *Oxidation in Food*, A.J. St. Angelo, Ed., American Chemical Society, Washington, D.C., 1992, p. 266.

22. B.S. Mistry, T. Reineccius, and L.K. Olson, GC-O for the determination of key odorants in foods, *Techniques for Analyzing Food Aroma*, R. Marsili, Ed., Marcel Dekker, New York, 1997, p. 265.

23. T.H. Parliment, Applications of a microextraction class separation technique to the analysis of complex flavor mixtures, *Flavor Analysis: Developments in Isolation and Characterization*, Washington, D.C., 1998, p. 8.

24. M. Chien and T. Peppard, Use of statistical methods to better understand gas chromatographic data obtained from complex flavor systems, *Flavor Measurement*, C.-T. Ho and C. Manley, Eds., Marcel Dekker, New York, 1993, p. 1.

25. R.T. Marsili and N. Miller, Determination of the cause of off-flavors in milk by dynamic headspace GC-MS and multivariate data analysis, *Food Flavors: Formation, Analysis and Packaging Influences*, E.T. Contis, C.-T. Ho, C.J. Mussinan, T.H. Parliment, F. Shahidi, and A.M. Spanier, Eds., Elsevier Science B.V., The Netherlands, 1998, p. 159.

26. R.T. Marsili, SPME-MS-MVA as a rapid technique for assessing oxidation off-flavors in foods, *Headspace Analysis of Foods and Flavors: Theory and Practice/ Advances in Experimental Medicine and Biology*, Vol. 488, R.L. Rouseff and K.R. Cadwallader, Eds., Kluwer Academic/Plenum Publishers, New York, 2001, p. 56.

27. R.T. Marsili, Combining mass spectrometry and multivariate analysis to make a reliable and versatile electronic nose, *Flavor, Fragrance and Odor Analysis*, R. Marsili, Ed., Marcel Dekker, New York, 2002, p. 349.

28. R.T. Marsili, Shelf-life prediction of processed milk by solid-phase microextraction, mass spectrometry, and multivariate analysis, *J. Agric. Food Chem.* 48: 3470, 2000.

29. B. Vallejo-Cordoba and S. Nakai, Keeping-quality assessment of pasteurized milk by multivariate analysis of dynamic headspace gas chromatographic data. 1. Shelf-life prediction by principal component regression, *J. Agric. Food Chem.* 42: 989, 1994.

30. K.R. Beebe, R.J. Pell, and M.B. Seaholtz, *Chemometrics: A Practical Guide*, John Wiley & Sons, New York, 1998, p. 63.

31. F. Drawert and N. Christoph, Significance of the sniffing technique for the determination of odor thresholds and detection of aroma impacts of trace volatiles, *Analysis of Volatiles: Methods and Applications*, P. Schreier, Ed., De Gruyter, Berlin, 1984, p. 120.

32. W. Grosch, Determination of potent odourants in foods by aroma extract dilution analysis (AEDA) and calculation of odour activity values (OAVs), *Flav. Fragr. J.* 9: 147, 1994.

33. M. Preininger and W. Grosch, Evaluation of key odorants of neutral volatiles of Emmentaler cheese by the calculation of odour activity values, *Lebensm. Wiss. Technol.* 27: 237, 1994.

34. R.J. McGorrin and L. Gimelfarb, Comparison of flavor components in fresh and cooked tomatillo with red plum tomato, *Food Flavors: Formation, Analysis and Packaging Influences*, E.T. Contis, C.-T. Ho, C.J. Mussinan, T.H. Parliment, F. Shahidi, and A.M. Spanier, Eds., Elsevier Science B.V., The Netherlands, 1998, p. 295.

35. J.E.R. Frijters, Some psychophysical notes on the use of the odour unit number, *Progress in Flavour Research*, D.G. Land and H.E. Nursten, Eds., Applied Science, London, 1979, p. 155.

36. M. Qian and G.A. Reineccius, Quantification of aroma components in Parmigiano Reggiano cheese by a dynamic headspace gas chromatographic-mass spectrometry technique and calculation of odor activity values, *J. Dairy Sci.* 86: 770, 2003.

37. T.E. Acree, J. Barnard, and D.G. Cunningham, A procedure for the sensory analysis of gas chromatographic effluents, *Food Chem.* 14: 273, 1984.

38. R.T. Marsili and N. Miller, Determination of major aroma impact compounds in fermented cucumbers by solid-phase microextraction-gas chromatography-mass spectrometry-olfactometry detection, *J. Chromatogr. Sci.* 38: 307, 2000.

39. P. Pollien, A. Ott, F. Monigon, M. Baumgartner, R. Munoz-Box, and A. Chaintreau, Hyphenated headspace-gas chromatography-sniffing technique: screening impact odorants and quantitative aromagram comparisons. *J. Agric. Food Chem.* 45: 2630–37, 1997.

40. A. Ott, L.B. Fay, and A. Chaintreau, Determination and origin of the aroma impact compounds of yogurt flavor, *J. Agric. Food Chem.* 45: 850, 1997.

41. R.T. Marsili, N. Miller, G.J. Kilmer, and R.E. Simmons, Identification and quantitation of the primary chemicals responsible for the characteristic malodor of beet sugar by purge and trap GC-MS-OD techniques, *J. Chromatogr. Sci.* 32: 165, 1994.

42. R.T. Marsili, Application of SPME GC-MS for flavor analysis of cheese-based products, presented at the American Chemical Society 228th National Meeting and Exposition, Philadelphia, August 22–26, 2004, in press.

43. W. Engel, W. Bahr, and P. Schieberle, Solvent assisted flavour evaporation — a new and versatile technique for the careful and direct isolation of aroma compounds from complex food matrices, *Eur. Food Res. Technol.* 209: 237, 1999.

44. P. Werkhoff, S. Brennecke, W. Bretschneider, and H.-J. Betram, Modern methods for isolating and quantifying volatile flavor and fragrance compounds, *Flavor, Fragrance and Odor Analysis*, R. Marsili, Ed., Marcel Dekker, New York, 2002, p. 139.

chapter four

An integrated MDGC-MS-olfactometry approach to aroma and flavor analysis

David K. Eaton, Lawrence T. Nielsen, and Donald W. Wright

Contents

I. Introduction

Recent years have seen the emergence of a number of innovative technologies for the investigation of aroma and flavor-quality issues for food, beverage, and consumer products. These have included, among others, e-nose [1], ms-nose [2], fast-GC [3], ion trap GC-MS [4], and 2D-GC-TOF [5]. Although each of these technologies has demonstrated some utility for limited applications, in our experience none of these has been as directly or widely effective as integrated multidimensional gas chromatography-mass spectrometry-olfactometry (i.e., MDGC-MS-O) for rapidly pinpointing and resolving problems related to aroma and flavor quality [6].

The investigation of aroma and flavor quality problems related to food and beverage products represents a unique challenge for the analytical chemist [7]. The typical response to aroma and flavor correlation challenges is to extract or concentrate the volatile components from the product headspace, separate these components by GC, identify them by mass spectrometry, and then speculate as to which of the scores (or perhaps hundreds) of compounds account for the aroma attributes of interest. This approach is seldom successful due to the fact that the most important contributors to aroma and flavor quality are typically ultratrace compounds buried in a forest of volatiles which are insignificant with respect to aroma and flavor. It is this limitation which makes MDGC-MS-O such a uniquely appropriate approach for this analytical challenge [7,8,9,10]. MDGC techniques of heart-cutting, cryogenic trapping, and back-flushing enable needle-in-the-haystack, critical flavorant separation from the complex background [11,12]. Concurrently, the integrated olfactometry and mass selective detectors enable aroma impact assessment (i.e., character and intensity) and chemical identification for the compounds separated in this manner. This data, when considered in relation to the composite odor character of the sample, will typically enable the development of correlations between critical odorants and their impact on composite odor quality.

This integrated approach is clearly based upon well-established individual technology elements. These elements include gas chromatography — 1951 [13], open tubular columns — 1957 [14], Dean switch-based MDGC — 1968 [8], and GC-olfactometry — 1964 [15]. However, it is largely the result of the subsequent refinements to each of these elements that are responsible for the routine utility of the integrated technology as it stands today. Critical refinements have included, but are not limited to, fused silica capillary technology, mass selective detector development, cryogenic trapping capabilities, and hardware miniaturization, along with system passivation, control, automation, data-acquisition, and processing. Each of these areas of refinement are heavily reflected in state-of-the-art integrated MDGC-MS-O systems.

In the experience of these authors, an MDGC-MS-O approach has proven to be particularly effective in enabling a rapid response to crisis-driven aroma and flavor quality problems in the food, beverage, and associated packaging industries. In addition to problem solving, this approach has proven equally effective for prioritizing the critical aroma components in baseline studies of "gold standard" reference samples for new or established products. Such aroma profile information can be invaluable with respect to subsequent routine production tasks, enabling a rapid response to aroma quality deviations from the predefined gold standard. In addition to such preemptive problem "interception," this data can be used in on-going efforts to improve the aroma and flavor characteristics of established food or beverage products.

The integrated MDGC-MS-O approach presented here has evolved out of a requirement for maximum efficiency and minimum response time to crisis-driven aroma problems. In the three sections which follow, this MDGC-MS-O based investigative strategy is illustrated using application examples. These applications explore three distinct aspects of the technology: malodor and malflavor correlation studies, integrated aroma/flavor quality control strategies, and application-driven system design. Referenced application areas include residual odor volatiles in packaging, general aroma profiling by GC-O, and the exploration of links between aroma volatiles and flavor quality.

II. Malodor/malflavor investigations

Experience has shown that aroma and flavor defects can originate from a remarkably wide variety of sources [16]. These sources include changes internal to the product itself, such as a flavor change brought about by permeation loss, or reactive conversion [17] of key flavor components during storage. Likewise, in many cases, these defects have also been shown to result from factors external to the product itself. These external factors include cross-contamination brought about by contact with contaminated environments or packaging materials [18,19] such as containers, closures, labels, adhesives, and printing inks. Potential environmental cross-contamination sources include all environments with which the product is brought into contact, including production, storage, shipment, retail, or consumer residence.

Malodor issues are very often context-driven. For example, although the strong "vomitus" and "body odor" character of butyric and isovaleric acids, respectively, are very offensive in social situations, they are surprisingly prominent aroma contributors to many of the cheese-based snack foods we savor.

Likewise, a distinct floral aroma character which would be perceived as very pleasant when encountered in a garden setting would, almost certainly, be perceived as an aroma defect in bottled drinking water. In fact, one of the most severe challenges to plastic packaging is bottled drinking water. Water, being aroma-neutral, has limited masking ability, and as a result, any odor character which is detectable by the consumer will be perceived as a flavor-quality defect. Therefore, the packaging of drinking water in plastic containers places special constraints on the quality of packaging materials, as well as all other potential sources of odor cross-contamination. In practice, odor defects, representing all of the possible sources alluded to in the previous section, have been encountered relative to drinking water packaged in plastic containers.

Regardless of the type of product or environment which is the focus of a malodor investigation, the aroma-profiling process is basically the same. This process involves the following major tasks:

1. Collecting, concentrating, and utilizing GC-O to develop an aroma profile of the equilibrated volatiles within the contained headspace of a representative worst-case sample.
2. Developing an odorant priority ranking profile relative to odor intensity and character.
3. Correlating GC-O odor profile priority rankings with sensory panel gradations of the sample under investigation.
4. Utilizing the odorant-priority ranking information from the worst-case sample to perform odor profile cross-comparisons to representative best-case equivalents. It is often the case that this step will yield first stage confirmation of an initial hypothesis regarding odor-defect carrier compounds.
5. Focusing on the suspect "bad actor(s)" emerging from the previous investigative steps, perform targeted cross-comparisons relative to materials or processes associated with the defective product. Establishing a process, raw material, or environmental link with the suspect malodorants will often reveal the primary source of an odor-quality excursion.

The general sampling strategy relative to malodor/malflavor investigations is to process the sample under experimental conditions which most closely match the conditions associated with the consumer complaint. For example, if the complaint is one of malodor, the product headspace will typically be sampled directly rather than taking a less direct liquid extraction approach. Likewise, if an aroma defect is perceptible to the consumer at room temperature, the initial headspace sampling work will also be carried out at room temperature rather than by sampling at elevated temperatures. However, it is often the case that some condition of production, packaging, storage, preparation, or end use is required before an aroma or flavor defect becomes detectable in the finished product. Such conditions include heat stressing, "flooding out," "salting out," or precursor oxidation. In those cases, it is often necessary to adopt sample preparation strategies which match those critical conditions before the defect will become evident. The case studies presented in this section demonstrate how these investigative strategies can be used to resolve even complex aroma-defect introduction pathways.

A.   Case study #1 — precursor oxidation in bottled drinking water

The general malodor investigative process is illustrated in the first case study summarized below. This project involved an off-quality production lot of bottled drinking water which was packaged in polyethylene terephthalate (i.e., PET) containers. In this case, the malodor descriptor for the product was floral or herbaceous, not unpleasant in a different context but a clear odor-quality defect for bottled drinking water. The mass spectrometric-TIC profile shown in Figure 4.1 reflects the overall complexity of the equilibrated volatiles, which were collected from the vapor headspace formed inside one of these containers. In contrast, the corresponding aromagram (Figure 4.2), which was developed in parallel by GC-O, was shown to be much simpler as only a few significant odor notes were detectable under the initial test conditions. More importantly, it was determined that the most prominent of the detected odor notes appeared to be mainly responsible for the composite odor defect. This was indicated by a strong similarity in odor character when comparisons were made between the composite odor of the product itself and that of the isolated odor note. This strong similarity indicated that this single compound was character-defining relative to this particular odor defect. At this point the focus shifted quickly from characterizing and prioritizing the critical odorant to identification and source determination.

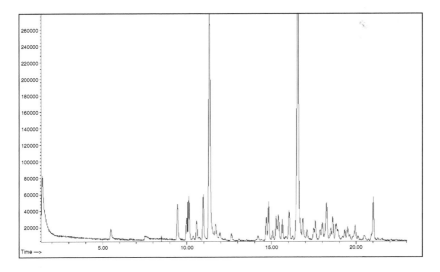

Figure 4.1 Headspace volatiles MS-TIC trace — PET packaged bottled water — high odor.

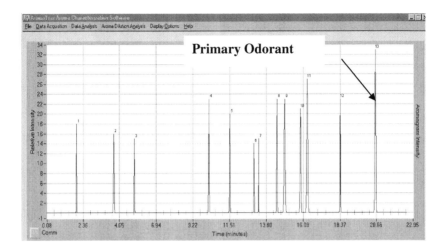

Figure 4.2 Headspace volatiles aromagram — PET packaged bottled water —— high odor.

As is often the case, it was not possible to develop direct mass spectrometric identification of the critical odorant due to its extremely low concentration level and the presence of high-level, coeluting interference peaks.

However, as shown in Figure 4.3 through Figure 4.5, it was possible to use MDGC techniques of heart-cutting and cryotrapping to isolate, quantify,

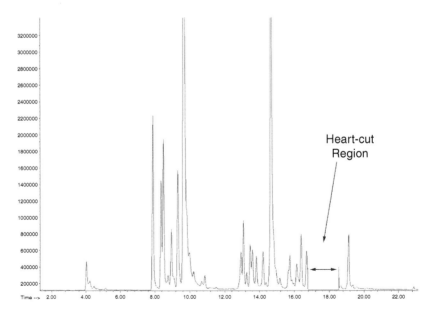

Figure 4.3 Precolumn separation minus heart-cut — PET packaged bottled water — high odor.

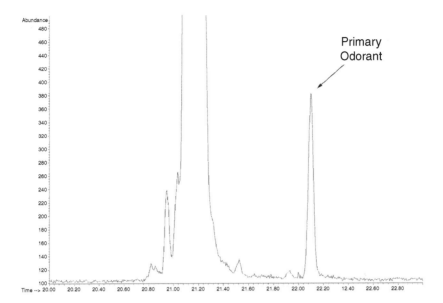

Figure 4.4 Primary odorant — PET packaged bottled water — Process 1 — high odor.

identify and track the critical odorant within the product and bottling process. The MDGC-based separation of the critical region enabled isolation of the primary malodorant from the interfering background. This separation, in turn, enabled a clean MS spectra to be developed and tentative identification

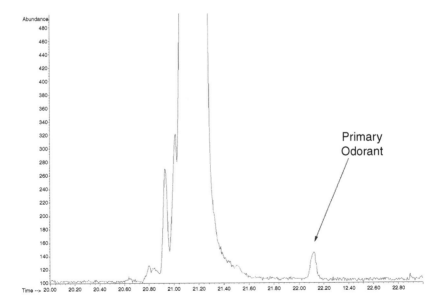

Figure 4.5 Primary odorant — PET packaged bottled water — Process 2 — low odor.

of the critical odorant as a C-11 aldehyde. The final step in malodor-source determination was to take this information and profile the critical odorant within the bottling process. To this end, a process trial was carried out which involved operating the process under two different sets of conditions during which all other parameters such as packaging, water supply, etc., were held constant. As shown in Figure 4.4 and Figure 4.5, one of the process variables was shown to result consistently in the production of a 5- to 7-fold higher concentration of the critical odorant. This increase in concentration occurred in concert with a corresponding down-grading of the product odor and flavor quality from acceptable to unacceptable. Interestingly, even though this odor defect was shown to be the result of an oxidation process, the expected through-wall migration of oxygen did not appear to be a significant contributor. Instead, an indirect route appeared to be responsible: A volatile precursor from one of the packaging components was oxidized within the process and converted into the malodorant ultimately responsible for the flavor-quality defect. As is often the case, the malodorant precursor itself did not carry a significant odor impact. The ultimate solution to this problem was the elimination of the precursor from the packaging component as the oxidizing process variable was an essential part of the pasteurization process.

B. Case study #2 — aroma/flavor defect in plastic-wrapped cracker product

A common misconception encountered relative to malodor issues is the notion that a positive chemical identification of a critical odorant is absolutely required to pinpoint and correct an aroma-quality defect. Certainly, it is always a goal to identify these problem compounds, whenever possible. In practice, however, quality defects resulting from even very complex introduction pathways are often resolved without achieving positive identification of the responsible compounds. Factors commonly acting as barriers to compound identification include the target-compound spectra not represented in available spectral libraries, and representative defective samples that are mass-limited relative to the extreme potency and trace concentration levels of the critical odorants. However, the overriding consideration is often simply the additional effort and associated expense required to extract, concentrate, isolate, and identify a high-impact, trace-level odorant from a complex matrix. Each sensory-defect question must be evaluated individually and a decision made as to whether malodorant identification is truly essential or simply a curiosity, especially if the source of the immediate problem can be readily identified and the defect corrected without the added expense of odorant identification. The second case study which is presented below serves as a good example of such a circumstance.

This project was focused on a crisis-driven flavor defect which emerged relative to a high-volume, plastic-wrapped cracker product. In contrast to the out-of-context floral defect previously described for bottled drinking water, this cracker product was marked by a very unpleasant bleach-like odor and flavor defect. The critical odorant which was responsible for this defect had eluded discovery by conventional GC-MS based approaches for several months before we became involved in the investigation. The unique appropriateness of an MDGC-MS-O approach to resolving problems such as this is indicated by the fact that the critical odorant was located within the first two-odor profile screening runs subsequent to beginning the GC-O phase of investigation. As is often the case, this flavor defect was shown to be carried by a single odorant of high-odor potency which was present in the sample headspace at extremely minute levels. The "character-defining" impact of this compound was indicated by the fact that the odor character of the chromatographically separated compound and that of the defective product itself were virtually identical. At this point the focus shifted quickly from defining the critical odorant(s) to MDGC-based compound isolation and identification. As is often the case, it was not possible to develop a clean mass spectral profile for the critical odorant without this MDGC-based isolation. This inability resulted from the combined effects of the low concentration level of the critical odorant, and the complexity of the product headspace emissions.

Figure 4.6, below, shows the precolumn separation minus the target heart-cut region, which was found to contain the malodorant. As shown in this chromatogram, the trace-level odorant was buried in a region of tremendous complexity which had to be reduced before the compound could be isolated and identified.

Figure 4.7 and Figure 4.8, in turn, present the relative responses for the target odorant when equivalent heart-cut analyses were carried out for the

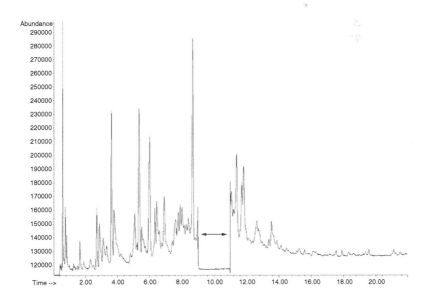

Figure 4.6 Precolumn separation scan minus isolating heart-cut region.

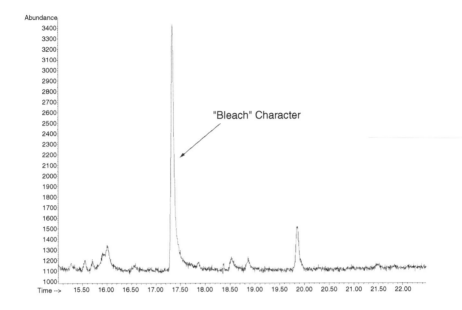

Figure 4.7 MS-SIM scan of isolating heart-cut region for worst-case sample.

representative worst-case and best-case samples. The comparative MS responses in combination with parallel olfactory responses are taken as solid first-stage confirmation of the single compound source of the odor or flavor defect.

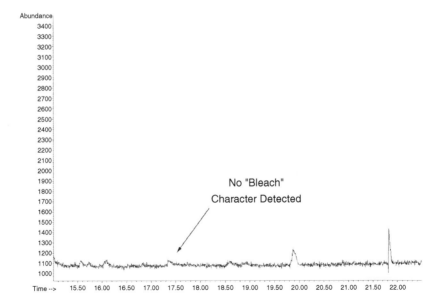

Figure 4.8 MS-SIM scan of isolating heart-cut region for best-case sample.

In this case, it was quickly determined that, in spite of the very clean mass spectral profile (i.e., enabled by heart-cut isolation) there was no published spectral library match found for this compound. However, in spite of the lack of chemical identification of the odorant, possession of good mass spectral profile data enabled an MS-selected ion monitoring (i.e., MS-SIM) method to be quickly developed and used in determining the pathway by which this odorant was being introduced into the product. The subsequent target-odorant tracking study was carried out in step-wise fashion. Step 1 confirmed that the target odorant was common to the printed packaging film, but not to the cracker product which had not been exposed to the packaging film. Step 2 confirmed that the target odorant was common to printed film, but not to the equivalent unprinted film. Step 3 confirmed that the target odorant was common to only one of the seven inking components which were used in printing the film. The problematic inking component was found to be a carrier solution which was used for transporting the colorants during the printing process. Subsequent to correlating this odor defect with its source, it was determined that the problematic printing component was supplied by one vendor. In turn, research efforts carried out by this vendor ultimately confirmed that the appearance of the odor defect in its product corresponded to a change that was made in the formulation of the product. The efficacy of the MDGC-MS-O approach to these types of crisis-driven flavor-quality excursions is clearly demonstrated through this application. Remarkably, only three working days were required to achieve complete resolution of a flavor-defect issue which had eluded resolution by other approaches for several months. Clearly, pressing on with additional efforts aimed at chemical identification of the responsible malodorant would have represented a significant additional effort and expense and would have served no useful purpose with respect to resolving this quality defect. Obviously, based upon this consideration, the decision was made not to pursue chemical identification of the responsible odorous compound. Its chemical identity remains unknown to us to this day.

C. Case study #3 — malflavor in powdered smoked meat flavor concentrate

Case study #3 was focused on a high-priority flavor defect relative to a "smoked meat" flavorant which was supplied to snack food producers in a dry-powder concentrate form. When applied to a potato chip snack product the defective lot of this concentrate resulted in an unusual flavor defect; one which was best characterized as sweet and fruity. This case study represents another interesting example of the context-dependent nature of flavor-defect issues. In this case, the flavor character of the defective product was, in fact, pleasant when considered on its own merit. Unfortunately, when considered within the context of the targeted-flavor character for this particular product, a salted, smoked meat-flavored potato chip, its flavor had to be graded as uncharacteristic and unacceptable.

As stated previously, the general sampling strategy relative to malodor investigations is to process the sample under experimental conditions that most closely match the conditions associated with the consumer complaint. This consideration can be particularly important in the case of flavor-defect issues relative to low moisture-content food products such as the fried potato chips under current consideration. In such cases, it is often advantageous to collect the headspace volatiles both before and after saturation of the dry product with water or normal saline solution. These are extremely useful sample preparation techniques which are often referred to as flooding out and salting out. It should not be surprising that the profiles of chromatographic volatiles (i.e., and corresponding aroma profiles) are often dramatically different when direct comparisons are made between dry and wet conditions. More importantly, the addition of moisture to a dry food product establishes conditions, relative to the sample, that more closely reflect the condition which develops in the mouth when the product is chewed. In practice, it is often the case that a critical aroma difference between defective and control samples only becomes readily apparent after the initiation of a flooding out or salting out effect. Such is the case regarding the defective-flavored potato chip product under current consideration.

The two aromagrams shown in Figure 4.9 and Figure 4.10 below reflect the significant differences in aroma profiles which were developed for the same sample under dry and wet conditions. In this case, all test parameters were held constant (i.e., SPME fiber, sample, headspace vessel, equilibration temperature, and collection time) except for the addition of water to saturation prior to the second collection. A note by note comparison of these two aromagrams shows a significant number of relative response shifts in the aroma profiles. Particularly noteworthy in the "flooded out" profile are relative response increases in the buttery notes carried by diacetyl (peak #5) and 2,3-pentanedione (peak #8), as well as the fruity notes due to ethyl butyrate (peak #9), ethyl 2-methyl butyrate (peak #11), and ethyl 3-methyl butyrate (peak #12).

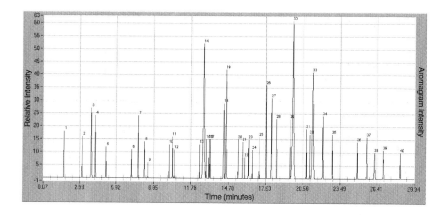

Figure 4.9 Aromagram profile of defective flavorant headspace before flood out.

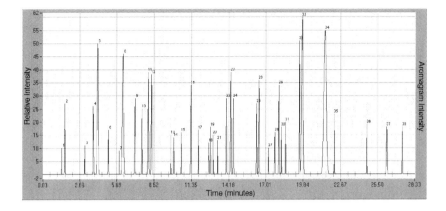

Figure 4.10 Aromagram profile of defective flavorant headspace after flood out.

Armed with the aroma-profile information for the defective flavorant lot under flooded out conditions, equivalent aroma-profile comparison runs were subsequently carried out for the control sample. As shown in Figure 4.11, the control flavorant-profile presented with significantly lower relative olfactory response intensities for the fruity butyrate ester series (peaks #7, #9, and #10) in comparison to the previous defective lot profile.

Correlating the qualitative aroma-profile differences between these two samples with corresponding MS-TIC response data revealed the following significant headspace concentration level differences between the flavor-defective and control-product samples (see Table 4.1).

The headspace concentration levels for these four flavorants were found to be significantly and consistently higher in the defective sample relative to the control. This was especially true relative to ethyl butyrate and ethyl 2-methyl butyrate, with corresponding higher headspace concentration levels

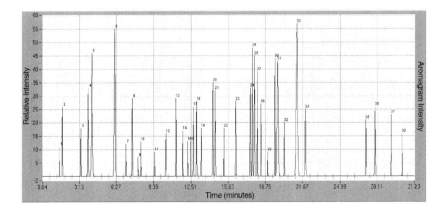

Figure 4.11 Aromagram profile of control flavorant headspace after flood out.

Table 4.1 Significant Headspace Odorant Differences — Defective vs. Control

Flavorant	Peak area response ratios (Defective/Control)
2,3-Pentanedione	1.3
Ethyl butyrate	7.0
Ethyl 2-methyl butyrate	11.0
Ethyl 3-methyl butyrate	1.5

of 7-fold and 11-fold, respectively, in the flavor-defective sample. These results were taken as first-stage confirmation with respect to the identification of the primary carriers of the flavor defect for the unacceptable lot. Follow-up discussion, with the supplier of the flavor intermediates used for formulating this product, indicated that there were suspicions of a possible cross-contamination incident relative to their batch of supplied materials. Specifically, there was the possibility of cross-contamination with a strawberry-flavorant which was being processed in their facility within the same time-frame. The GC-O based conclusions stated above were found to be consistent with the possibility of this strawberry-flavorant cross-contamination scenario. This was supported by published reference data [20] which listed ethyl 2-methyl butyrate as one of the primary, character-defining odorants present in strawberry juice. As was shown in the previous two case studies, the speed of response to such high-priority flavor-defect issues is made possible by the application of the integrated MDGC-MS-Olfactometry investigative approach. In this case, the initial hypothesis regarding the critical impact of the butyrate esters relative to the fruity, sweet flavor defect was developed within 2 hours of initiating the GC-O based investigation.

As demonstrated above, it is the ability to deliver rapid resolution of complex aroma and flavor-defect problems which has shown the integrated MDGC-MS-O approach to be uniquely suited for off-aroma and off-flavor investigations. However, the same qualities which make these techniques suitable for crisis-driven troubleshooting tasks have proven to be equally effective for slower-paced, general aroma profile work aimed at aroma-impact prioritization, flavor optimization and flavor-quality control. The above smoked-meat flavorant study represents an interesting example of this value. In spite of the brief, troubleshooting nature of the study a considerable amount of general flavor-correlation information was developed coincidentally. The following is a listing of some of the critical individual flavor components as collected from the equilibrated flooded-out headspace of this flavorant. These prominent and character-defining odorants are listed in an approximate descending order relative to their initially perceived significance to the composite-aroma character of the product (see Table 4.2).

The above listing represents only a first-pass approximation of the prioritized odorants relative to this flavorant. However, in spite of the cursory nature of this effort, a reasonable level of certainty emerged relative to the three to four compounds carrying the greatest individual-flavor impact. These include guiacol as the primary carrier of the characteristic smoky

Table 4.2 Priority Odorants — Smoked Meat Flavorant

Odorant	Descriptor
Guiacol	Smoky, medicinal, phenolic
2-Methylthio-3-methylpyrazine	Characteristic, savory
Trimethyl pyrazine	Savory, characteristic
Unknown @ RT 19.5 min	Cereal, hay, vitamin
Diacetyl	Buttery
Isovaleric acid	Body odor, musty
Unknown @ RT 18.0 min	Spicy, herbaceous
Methyl furfural	Nutty, musty
2,3-Pentanedione	Buttery, sweet
3-Methyl butanal	Foul, stale vegetable
Methional	Potato
Hexanal	Grassy

flavor and the two potent pyrazines as primary carriers of high-impact, savory-flavor character. An expanded GC-O-based aroma profile effort, carried out in conjunction with detailed sensory-panel correlations, would enable this first effort to be expanded and refined to a much higher level of definition and certainty. In the following section, such an expanded investigative effort is presented, relative to another complex flavor-quality challenge — the correlation of critical aroma volatiles and beer flavor.

III.   GC-O in aroma–flavor quality control

The first three case studies in this chapter show how GC-O is traditionally used for food and food-packaging analysis to identify the compounds that are responsible for off-flavors. Another important use of GC-O is for the development of routine flavor-quality control methods. When GC-O information is sufficiently accurate, it can be used as the basis for quality control applications that directly monitor the aroma compounds that are critical to a product's flavor quality. The success of this kind of QC method depends on how well the set of monitored flavor compounds is chosen, and that depends on how well the individual flavor characters are understood to relate to the whole sample flavor. The following two case studies illustrate how GC-O techniques impact flavor QC methodology.

A.   Case study #4 — vanilla-extract flavor quality: A univariate approach

The GC-O aroma profile of a vanilla extract will typically have as many as 40 significant aroma notes, depending on how the sample is collected and analyzed. Vanillin gives the principal flavor in vanilla extracts, and its level is one measure of quality. There are, however, numerous other flavor compounds present such as ethyl esters, anisaldehyde, methyl cinnamate, 4-vinylguaiacol, and several other high-impact flavor compounds, and their concentrations also impact flavor quality. Pure vanilla extracts are produced in many different grades of quality for different flavor applications. The type of bean, curing method, and extraction conditions are some of the major factors that determine the quality of the final product. The aroma profile contains information about aroma intensity and character of the individual compounds that make up the extract. This information can be used to evaluate the quality of extracts by comparing GC-O results of different samples and matching this with quality evaluations made by a flavor panel. To illustrate how one goes about obtaining information for the design of a vanilla-extract flavor QC method from GC-O data, one aspect of vanilla will be considered, the top-note, fruity flavor volatiles.

1. Designing a QC method

Top note fruity esters are not found in cured vanilla beans but are formed during the ethanol extraction process. The first 15 most volatile aroma compounds in an extract obtained at relatively high extraction temperatures are identified in Table 4.3. The relative intensities of the compounds in the headspace at ambient conditions were determined by the aroma dilution technique [21,22]. The headspace volatiles were collected by SPME, thermally desorbed into the GC, and split at the inlet by the ratios shown in the Table 4.3. Each flavor compound was detected at the GC-O sniff port up to its dilution ratio shown in Table 4.3. GC-O comparison of extracts produced under different extraction conditions gives different aroma-dilution plots and thus different intensity rankings of individual flavor components. The compound that is seen to vary most markedly with the process temperature conditions is ethyl acrylate. It is the dominant top note with high temperature extraction but is a minor flavor contributor in low temperature processing.

Not only is ethyl acrylate seen to have a significant relative concentration increase with temperature, but the flavor panel relates the product flavor

Table 4.3 Top-Note Aroma Compounds in High Temperature Vanilla Extracts

Compound (Order of increasing retention time)	Dilution ratio
Acetadehyde	9:1
Ethyl formate	30:1
1,1-Ethoxymethoxyethane + diethoxymethane	30:1
Ethyl acetate	90:1
Acetal	90:1
2-Methylbutanal + 3-methylbutanal	50:1
Ethanol	90:1
1,1-Diethoxypropane	9:1
Ethyl 2-methylpropionate	50:1
Ethyl acrylate	150:1
Ethyl butyrate	3:1
Ethyl 2-methylbutyrate	50:1
Ethyl isovalerate	30:1

change to the flavor character of pure ethyl acrylate diluted to the appropriate concentration. The flavor panel describes the whole extract as changing from sweet fruity to a harsher, dried-fruity flavor with higher extraction temperatures, a result consistent with the observed high ethyl acrylate levels. Low-temperature extraction produces less than 50 ppb ethyl acrylate from all types of vanilla beans. High-temperature extraction can produce over 1 ppm ethyl acrylate, depending on the conditions. Such high levels negatively affect flavor quality, but moderate levels are considered desirable for certain cold-served product applications.

2.   Univariate vs. multivariate data analysis

A vanilla-extract flavor quality method starts with information taken from aromagrams and aroma-dilution plots. In the case of high temperature vanilla extraction, it may be necessary to monitor only ethyl acrylate for product-flavor quality even though many other compounds contribute to the overall flavor. In this case, QC can be done by a classical univariate technique such as GC-FID.

If a product has many aromas that impact flavor quality, then multivariate-based analysis (MVA) techniques are needed to handle the increase in the complexity of the data. Precise GC-O data permits the selection of the most important target-flavor compounds in complex samples. The next section shows how to develop this information and use it to build a complete QC flavor method using the mass spectrometer as detector and multivariate data analysis.

B.   Case study #5 — beer flavor quality: A multivariate approach

1.   Variability of beer flavor

Food products have variable composition and thus variable flavor. This is true because their ingredients are not generally uniform. For a product like beer, a wet growing year for barley will affect the flavor quality produced from that crop. Hops may come from different sources and vary in quality. They also change flavor on storage. Yeast will naturally change with time. Production parameters may not always be kept exactly the same. The number of variables involved in brewing is large, and the composition of the final product depends, to some degree, on all of them in a complex way. A producer would like to know how these factors impact quality so that corrections can be made to avoid problems with customer acceptance.

2.   Beer aroma comparison

GC-O analysis is the preferred technique to determine which aroma compounds are responsible for the aroma, and aroma changes in beer. With accurate aroma-profile data, a multivariate QC method can be quickly developed to monitor the key flavor-impact compounds. The GC-O analysis of a single beer product tells how many aroma compounds are present, their aroma character, and their intensity at the sniff port. What is then needed to build a QC method is to determine which of the many detected aroma compounds are related to the flavor differences between samples of different quality. The main problem with the direct GC-O approach is that beer samples, with all the possible variations of flavor, are often not available at the time of method development. The samples that are available are analyzed by GC-O techniques, and the compounds that are found to impact the flavor variations are incorporated into the QC method, but other means of identification must be found to complete the set of monitored compounds. They are described in the following section.

3. Importance of the flavor panel

Input from a flavor panel that has become familiar with a product is critical to help establish which aroma compounds need to be included in the QC method. A trained beer flavor panel will evaluate a given beer product against a control by numerically rating such flavors as hoppy, malty, fruity, floral, yeasty, skunky, etc. The panel collectively knows when a beer sample is out of specification and names the flavor defect that is the cause.

The analytical challenge is to link the flavor panel's flavor-character descriptions to the aroma descriptions sensed at the GC-O sniff port. For beer, the GC-O analyst may sense as many as 40 to 60 aroma components from a given sample and thus produce a complex aromagram. The aromagram in Figure 4.12 shows a typical GC-O beer analysis. Table 4.4 gives some of the corresponding aroma-character descriptions from the analysis. Some of the descriptors will match exactly with the flavor-panel flavor descriptions found in their evaluations of the whole sample. For example, the fresh hops' character matches exactly with the unique aroma of beta-myrcene. Likewise, when the flavor-panel finds skunky, the GC-O results will show corresponding sulfur-containing compounds that are responsible for the skunky flavor. In this way, a large number of correlations can be made between the flavor-panel descriptions and the GC-O aromagrams. Although many flavors can be

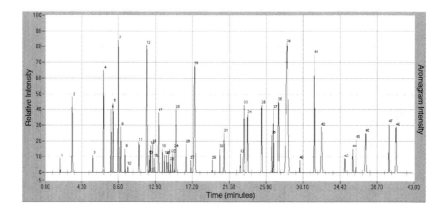

Figure 4.12 Typical aromagram of beer collected by SPME.

Table 4.4 Partial List of Aromas from Aromagram in Figure 4.12

Aroma number	Aroma description
7	Fusel oil
8	Banana
11	Fresh hops
15	Musty
17	Skunky
18	Aldehydic
19	Brothy
21	Roasted
23	Aldehydic
24	Nutty
25	Dill
27	Foul acid
31	Grainy
33	Hops
37	Rose
38	Dried fruity
39	Sweet

identified directly in this way, some must be further isolated and identified using the multidimensional GC techniques described in the earlier sections of this chapter. Those requiring MDGC are the high-flavor impact, trace-level compounds that are otherwise buried in the background of a standard GC-MS analysis. The nonvolatile beer flavors, such as bitter from the alpha-acids, are of course not addressed in the method developed from GC-O data.

4.   Indirect aroma correlations

Sometimes a panel-flavor description cannot be directly correlated to any single GC-O descriptor. To include such a panel-defined description in the QC method means that a judgment must be made as to what GC-O detected compound or compounds are associated with that panel description. These flavor characters are often made up of multiple flavors and give a flavor impression different from any single compound. The analyst must choose from among the aromas sensed at the sniff port those compounds most likely to make up the flavor that the panel has described. This often involves making an educated guess. The validity of the initially guessed flavor compounds is confirmed or rejected when the QC method is running, because their correlation or lack of correlation to the flavor panel's evaluations becomes evident in the data. In addition to the deletion of noncorrelating compounds, new aroma compounds can be added to fill gaps in the method that may have been overlooked during the initial method-development phase.

5.   Minimum aroma monitoring set

The ideal QC method for beer-flavor monitoring would measure concentration variations of only those aroma compounds that actually impact on the flavor quality, as sensed by human taste, and filter out all other compounds. The previous sections show how to find possible candidates for such a method through correlation of GC-O results with flavor-panel results. In the case of beer and many other types of samples, the number of these compounds can be relatively large. In order to design a method with a manageable number of monitored aromas, some compound selection is made from among the aroma candidates.

Reduction of the number of monitored compounds is performed to obtain a minimum-aroma test set. Each aroma is judged and those that appear to give redundant information are eliminated. An example of this would be to select only one of the many fruity ester compounds to monitor product fruitiness. A manageable set of compounds for the method might have a total of 10 to 15 compounds and would include compounds that monitor different aspects of the beer flavor. These would be markers to monitor the fermentation process, hops level, yeast integrity, selected high-impact trace level flavors and potential off-flavors. The object of the GC-O phase of the method development is to define the best possible set of aroma compounds that will monitor all the important variations that may occur in the brewing process, up to the point where the beer is sampled and analyzed. It may happen that some aroma compounds will have to be added to the method later when more detailed flavor information is discovered.

The process of determining which flavor compounds to monitor may take less than a week of GC-O work for less complicated products, but for more complex products like beer, where close collaboration with a flavor panel is required, more effort will often be needed. The time needed to finish the GC-O analysis depends on how much information is already known about the product. An extensive beer-flavor database will greatly aid in reducing the time for preparing a specific beer method.

6. Instrumental QC method

After a minimum-flavor test set is established, it is transferred to a routine analytical instrument. There are several possible instrumental approaches to using the GC-O data, but only one method will be described here. Because several of the beer components to be monitored are present at trace levels, the choice of instrument detector for the method is mass spectrometry operated in selected ion mode (SIM). This mass spectrometry technique detects a limited number of fragment ions keyed specifically to the aroma compounds in the test set. SIM detection works well when there is some degree of separation of the test compounds from extensive interference that is normally present in chemically complex samples such as beer. Introducing the entire collection of volatiles directly into the mass spectrometer, as is done in some MS-based QC instruments, will not generally give usable results with complex samples like beer. Therefore, a minimum chromatographic condition is found that effects the desired separation of the test set. Aroma compounds are thus detected using only selected ions in appropriate SIM chromatographic windows. The technique produces mass spectral data files

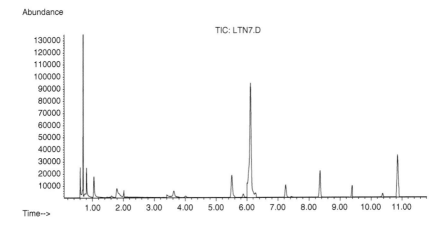

Figure 4.13 Minimum flavor test set GC-SIM chromatogram for flavor QC.

that contain mass intensity data in SIM windows that correspond to separated aroma-target compounds. This type of file is much simpler than a full mass spectra file for the same data collection time. As the total GC-MS run time for a minimum GC condition can normally be done for less than about 10 min, the method is in line with traditional QC analysis times. Figure 4.13 shows an example of such a GC-SIM chromatogram where a minimum flavor test set of 15 compounds is analyzed.

7. Headspace collection

For beer and many other products, headspace solid phase microextraction (SPME) is the sample collection method of choice. With SPME, no sample preparation is required. Beer is put into a headspace sampling vessel, temperature-equilibrated and sampled with an SPME fiber for a determined time. Reproducibility of this collection method depends on how well samples are equilibrated and how well the headspace collection time and fiber condition can be reproduced. When the same SPME fiber is used on each run, the analysis gives a reproducibility for most compounds in beer of better than 5% RSD. For trace-level compounds the reproducibility can be better than 15% RSD. For analysis that needs higher precision, other methods of collection are available at the cost of taking more time and effort.

8. Multivariate data analysis

The data generated by this QC method is treated by multivariate analysis [23–25]. Explained in simple terms, MVA takes data from multiple sample analyses and produces a vector made up of weighted contributions of each of the monitored compounds. The projection of the vector on to a two-dimensional plane gives a principal-component analysis (PCA) plot. Several dimensions are constructed mathematically, but only the first two or three axes are normally considered in the interpretation of the data. The reasons for using

MVA become clear when one tries to interpret the simultaneous variations of as many as 15 compounds by classical analysis. MVA takes all the data together from the 15 compounds and represents it as a point on a graph that is easy to interpret. Samples with similar composition will have vectors whose projected points cluster together on the PCA plot. As the samples diverge in composition, their vectors will fall in different positions on the plot. When statistically significant numbers of samples are analyzed, regions in the PCA plot are established that correlate to different product composi-tions and thus to different qualities. Newly analyzed unknown samples are compared with the plotted flavor patterns and are immediately classified as to their flavor quality by their position on the plot.

Figure 4.14 shows the PCA plot of the SIM data for 15 components of beer with each sample labeled by a flavor panel for their acceptability values. This numerical rating system is used for panel judging of beer that is being released for sale. The scale in this case is from 1 to 5 where 1 is target beer quality and 5 indicates an unacceptable product [26].

The MVA aroma database is a trained database established by running samples with the routine method, and by classifying them as to a flavor

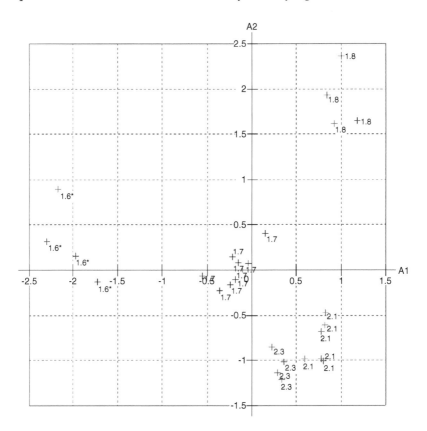

Figure 4.14 PCA plot of panel "acceptability" rating of beer with SIM flavor analysis.

panel's evaluation or other discriminating criteria such as production date or location. In Figure 4.14, each point on the PCA plot is a run labeled according to the panel's system of acceptability ratings. For other applications, samples may have labels as simple as "good" and "bad," and if it is possible to classify them with more descriptive labels, then this can be done so that the PCA plot for different beer samples might have "hoppy," "fruity," or "off-flavor" character. The MVA algorithm will cluster together samples with the same label classifications and maximize the discrimination among different label groups. A trained set of samples establishes the reference database against which subsequent routine unknown samples are compared. The database can be continuously improved as more examples of differently designated samples are identified by the flavor panel and added to it.

9. Sampling at fermentation

The described method can be applied in several ways to brewing. Monitoring of the fermentation process is given here as an example. Beer samples are drawn at certain well-defined points in the fermentation process. The fermentation composition is then related to flavor-panel quality evaluations. With a trained fermentation database established, routine sample analysis will then pick out fermentation runs that are different and therefore may have potential problems. Because the analysis is rapid, there is time to find a solution to correct the flavor problem. Off-flavor batches of beer are not frequent but are costly when they do occur. Calculations can be made to show if such methodology is cost effective for a particular application.

10. Flavor drift

When the same fermentation tanks are monitored over extended periods of time, the results will indicate how the aroma composition is changing during that time. These results measure the "flavor drift" in the product. The brewer may need to adjust the process to compensate for the flavor drift if it is great enough to show up in the flavor-panel evaluations. If the fermentation is not perfectly temperature controlled, then a flavor drift related to temperature might be measured from season to season. Long-term monitoring may show effects of raw material changes from one harvest to another, aging of yeast, changes in operating procedures or hardware changes. The method can establish a flavor baseline at some point in time and then monitor the product relative to the baseline standard. Long-term monitoring instructs operators how their system functions and indicates what to do if composition strays from the expected quality target. Figure 4.15 shows a PCA plot from commercial beer production for the flavor drift monitored by a set of 15 beer flavor compounds over 10 months measured by grouping each month together [26,27]. Only four of the 10 months are shown in the figure for simplicity. The plot shows a counterclockwise circular drift from the starting month of September (s) through December (d), February (f) and June (u).

The same method can be used to help reconcile the differences in flavor of the same beer made at two different production sites. For this application,

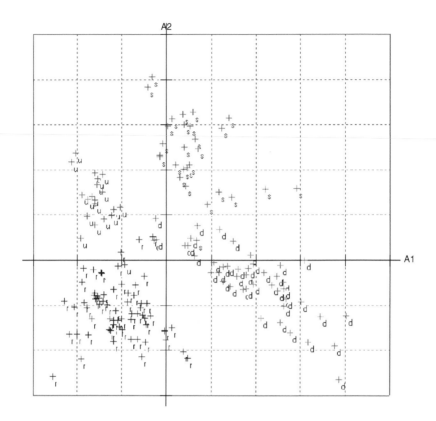

Figure 4.15 PCA plot of variation in beer flavor compounds.

the MVA baseline data consists of beer made at the "standard flavor" location. Comparison of the data from the locations gives detailed information about what compounds are responsible for the flavor difference. This information can be used to direct the correction effort to bring the two beers closer in flavor. The process is monitored by checking where on the PCA plot the flavor composition falls in relationship to the baseline after process changes are made.

IV. Integrated MDGC-MS-O system design strategy

Integrated MDGC-MS-O systems can be configured to reflect a wide range of complexity. This complexity can include multiple gas chromatographs, multiple ovens, multiple-selectable detectors, redundant detectors (i.e., such as multiple sniff ports) and complex stream-selection options. Although each of these options can add valuable flexibility relative to certain tasks, it is important to remember that increasing flexibility is almost always accompanied by increasing complexity. Simply stated, the more functions which are available for optimization, the greater the number of functions which will need to be accounted for (i.e., to some degree) in each analytical run. In designing an integrated MDGC-MS-O system, it is best, therefore, to design for a minimum level of complexity which can successfully address the majority of the day-to-day requirements in a particular laboratory situation. This conservative approach ensures that the increased complexity of an over-designed system does not become a constant burden relative to the more mundane daily separation tasks. The down-side of such a conservative approach is that there will likely be occasional separation challenges for which the increased complexity would be beneficial or required. This is especially true relative to aroma and flavor analysis due to the extreme headspace volatiles complexity of most food, beverage, and natural products. However, creative application of the functions of a basic MDGC-MS-O system can often be used to emulate the functions of a more complex integrated system. An example of such creative application is illustrated by Case Study #6, described later: the use of the cryotrap hold and oven-down program functions in a single GC system to emulate a dual-GC capability.

A common misconception relative to dual-column MDGC systems is that it is essential that the two columns be housed in independent, temperature-controlled zones such as separate GC ovens. While this coupling is rather straightforward to accomplish mechanically, the use of two GCs for a single MDGC system is a costly option, and much more mechanically complicated than a single oven alternative. In practice, there are three possibilities for achieving any heart-cut separation on a dual column/single oven MDGC system. These are:

- An on-the-fly heart-cut from the precolumn to the analytical column without intermediate cryotrapping at the head of the second column. If the two columns represent different phase selectivity, this option often yields the best separation.
- Heart-cut with cryotrapping at the head of the second column followed by immediate release from the cryotrap and separation on the second column.
- Heart-cut with cryotrapping at the head of the second column followed by oven temperature down-programming prior to release from the cryotrap and reinitiation of oven temperature up-programming and separation on the second column.

The last of these options is the most complex and, in practice, is the least often required to address commonly encountered separation challenges. However, when required, this approach does enable the emulation of a dual oven/dual GC system by a simpler, single-oven alternative.

A. Case study #6 — wildflower honey: Aroma character-defining compound isolation

Wildflower honey is among the most interesting and most complex natural food products for GC-O aroma-profile study due to the natural variations associated with the locations and environments where it is produced. Recent aroma-profile efforts with a locally produced Texas wildflower honey reflected this complexity and produced very interesting results with respect to the prioritized, character-defining aroma carriers. In spite of the wonderful aroma quality of this product when evaluated as a whole, the GC-O-based aroma profile revealed surprisingly prominent fecal and sulfurous contributions from the organic sulfides (i.e., methyl mercaptan through dimethyl trisulfide), vomitus, and stale contributions from the organic acids (i.e., acetic through valeric) and barnyard contribution from *para*-cresol and isomers. However, as is often the case, the negative odor impact of these compounds was overshadowed by several wonderful aromas which were described as sweet, floral, and characteristic. Based upon these aroma profile efforts, it was concluded that the latter, pleasant aromas were character-defining with respect to the composite aroma character of this product. In fact, of the approximate 75 discrete aroma notes profiled from the collected headspace of this product, one specific note appeared to particularly reflect the composite character of this product. However, as a result of the scores (i.e., or perhaps hundreds) of individual volatile compounds present in the concentrated headspace of this sample and the trace level of this target odorant, it was difficult to achieve compound identification or a sensory impact link utilizing a single oven MDGC approach. Figure 4.16 through Figure 4.18 illustrate the use of heart-cutting, cryotrapping, and oven temperature down programming to emulate a more complex dual-GC configuration and thereby achieve this particularly difficult separation.

Shown in these figures are the aromagram responses superimposed on the corresponding photoionization traces. In progressing from the noncryotrapped run to the cryotrapped with intermediate oven cool-down run, an increase in correlation certainty with respect to the large, character-defining aromagram peak is demonstrated. The former left open the possibility of three or four compounds as potential carriers of the key character-defining aroma. These included para-anisaldehyde, cinnamic aldehyde and a large, rounded peak for which there was no mass spectra library match. In contrast, the latter yielded a relatively clear match of the target aroma note with cinnamic aldehyde. More importantly, it also helped eliminate *para*-anisaldehyde and the large, unidentified compound as likely contributors with respect to the targeted aroma note. Clearly, in this case, the creative application of the basic MDGC functions of a relatively simple single oven MDGC system were able to successfully emulate the greater flexibility of a more complex dual GC alternative. Experience has shown that most separation challenges can be met utilizing the first two of the listed strategies. However, special cases may occasionally be encountered where independent oven-temperature profiles for the two columns would be advantageous. For those rare cases it is possible to make use of the third alternative strategy to emulate the greater flexibility of a dual-oven system.

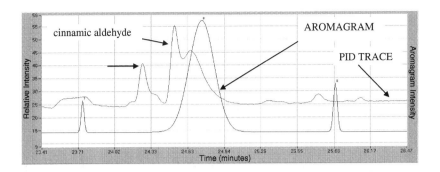

Figure 4.16 Mixed wildflower honey — heart-cut without cryotrapping.

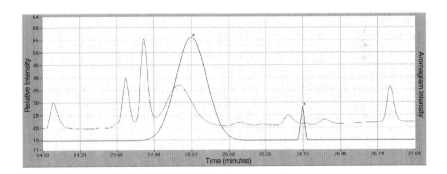

Figure 4.17 Mixed wildflower honey — cryotrap and release without oven cool-down.

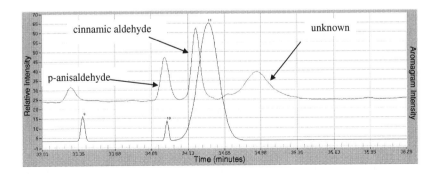

Figure 4.18 Mixed wildflower honey — cryotrap and release with oven cool-down.

V. Summary

As illustrated in the above series of case studies, an integrated MDGC-MS-O approach can be particularly effective for rapidly defining and prioritizing the basic building blocks of aroma and flavor character and quality. This integrated analytical approach has proven to be equally effective for both crisis-driven troubleshooting as well as baseline aroma profiling of product gold-standard samples. In contrast to other analytical technologies, MDGC-MS-O will typically yield very rapid correlations between targeted sensory attributes and the specific volatile compounds which are responsible for those attributes. Equally important is the fact that these sensory correlations can often be made based upon a minimal reference sample set. Such a minimal set may consist of only two samples: one sensory panel graded worst-case and an equivalent sensory panel graded best-case or control. Aroma profiling by MDGC-MS-O will typically indicate the primary aroma differences between these samples and, as illustrated, it is these differences which typically account for the corresponding differences in composite aroma/flavor quality. Such definitive information regarding specific compounds and corresponding key sensory attributes can, in turn, be used to develop targeted sensory quality control protocols. Depending upon the product, the resulting QC requirements may range from relatively simple, conventional GC-based univariate protocols to more complex GCMS-based multivariate alternatives. As illustrated in the above case study series, regardless of the required degree of protocol complexity, MDGC-MS-O can be a powerful tool for mapping the required protocol development.

Acknowledgments

The authors wish to express appreciation to Ray Marsili for his valuable counsel over the years and the opportunity to participate in this latest collaboration. We would also like to express appreciation to our many clients over the past decade whose collaboration and feedback have enabled us to refine and optimize our approach within this unique field of study.

References

1. D. Hodgins, The electronic nose: sensor array based instruments that emulate the human nose, *Techniques for Analyzing Food Aroma*, R. Marsili, Ed., Marcel Dekker, New York, 1997, p. 331.
2. R. Marsili, SPME-MS-MVA as an electronic nose for the study of off-flavors in milk, *J. Agric. Food Chem.*, 47: 2, 648, 1999.
3. D.W. Wright, D.K. Eaton, F.W. Kuhrt, and L.T. Nielsen, Evaluation of fast-GC and multivariate analysis for rapid odor quality monitoring of consumer products, *Proceedings of Pittsburgh Conference*, Chicago, IL, 2004.
4. C.K. Huston, Ion-trap mass spectrometry for food aroma analysis, *Techniques for Analyzing Food Aroma*, R. Marsili, Ed., Marcel Dekker, New York, 1997, p. 209.

5. J.M.D. Dimandja, S.B. Stanfill, J. Grainger, and D.G. Patterson, Application of comprehensive two-dimensional gas chromatography (GC × GC) to the qualitative analysis of essential oils, *J. High Resolution Chromatogr.* 23: 3, 208, 2000.

6. D.W. Wright, D.K. Eaton, F.W. Kuhrt, and L.T. Nielsen, MDGC-Olfactometry based investigations of odor quality issues in packaging and consumer products, *Proceedings of TAPPI Conference*, Indianapolis, IN, 2004.

7. D.W. Wright, Application of multidimensional gas chromatography techniques to aroma analysis, *Techniques for Analyzing Food Aroma*, R. Marsili, Ed., Marcel Dekker, New York, 1997, p.113.

8. D.R. Deans, A new technique in heartcutting in gas chromatography, *Chromatographia*, 1: 18, 1968.

9. G. Schomberg, H. Husmann, and E. Hubinger, Multidimensional capillary gas chromatography-enantiomeric separations of selected cuts using a chiral second column, *HRC&CC* 7: 404, 1984.

10. D.W. Wright, K.O. Mahler, and L.B. Ballard, The application of an expanded multidimensional GC system to complex fragrance evaluations, *J. Chromatogr. Sci.* 24: 60, 1986.

11. W. Bertsch, Methods in high resolution gas chromatography. 1. Two dimensional techniques, *HRC&CC* 1: 85, 1978.

12. B.M. Gordon, M.S. Uhrig, M.F. Borgerding, H.L. Chung, W.M. Coleman, J. Elder, J.A. Giles, D.S. Moore, C.E. Rix, and E.L. White, Analysis of flue-cured tobacco essential oil by hyphenated analytical techniques, *J. Chromatogr. Sci.* 26: 174, 1988.

13. A.P. Martin and A.T. James, *J. Biochem.* 50: 679, 1952.

14. M.J.E. Golay, *Gas Chromatography (1957 Lansing Symposium)*, V.J. Coates, H.J. Noebels, and I.S. Fagerson, Eds., Academic Press, New York, 1958, p. 1.

15. G.H. Fuller, R. Steltenkamp, and G.A. Tisserand, The gas chromatograph with human sensor: perfumer model, *Ann. NY Acad. Sci.* 116: 711, 1964.

16. B. Baigrie, Introduction, *Taints and Off-Flavors in Food*, B. Baigrie, Ed., Woodhead Publishing and CRC Press, Cambridge, 2003, p. 1.

17. R.J. Hamilton, Oxidative rancidity as a source of off-flavors, *Taints and Off-Flavors in Food*, B. Baigrie, Ed., Woodhead Publishing and CRC Press, Cambridge, 2003, p.140.

18. T. Lord, Packaging materials as a source of taints, *Taints and Off-Flavors in Food*, B. Baigrie, Ed., Woodhead Publishing and CRC Press, Cambridge, 2003, p.64.

19. P.A. Tice and C.P. Offen, Odors and taints from paperboard packaging, *TAPPI J.* 77: 149, 1994.

20. H.D. Belitz and W. Grosch, Fruits and fruit products, *Food Chemistry*, 2nd ed., Springer-Verlag, Berlin, 1999, p. 779.

21. T.E. Acree; Bioassays for flavor, *Flavor Science: Sensible Principles and Techniques*, T.E. Acree and R. Teranishi, Eds., American Chemical Society, Washington, D.C., 1993.

22. F. Ullrich and W. Grosch, Identifizierung den bei den Autoxidation von Linolsaure entstehenden Intensivsten Aromastoffe, *Z. Lebensm. Unters. Forsch.* 184, 277–282, 1987.

23. M.A. Sharaf, D.L. Illman, and B.R. Kowalski, *Chemometrics*, John Wiley & Sons, New York, 1986, p. 180.

24. C. Zervos and R.H. Albert, Chemometrics: the use of multivariate methods for the determination and characterization of off-flavors, *Off-flavors in Foods and Beverages*, G. Charalambous, Ed., Elsevier, Amsterdam, 1992, p. 669.

25. M. Chien and T. Peppard, Use of statistical methods to better understand gas chromatographic data obtained from complex flavor systems, *Flavor Measurement*, Chi-Tang Ho and C.H. Manley, Eds., Marcel Dekker, New York, 1993, p. 1.

26. D.K. Eaton, F.W. Kuhrt, L.T. Nielsen, and D.W. Wright, Flavor Quality Control, World Brewing Congress, San Diego, CA, 2004.

27. D.K. Eaton, L.T. Nielsen, and D.W. Wright, Applications of GC-O and MS-Nose Technologies for Quality Control in Brewery Operations, USDA Grant 2002-33610-12345, Final Project Report, 2004.

chapter five

Preseparation techniques in aroma analysis

Michael C. Qian, Helen M. Burbank, and Yuanyuan Wang

Contents

I. Introduction

Aroma analysis is highly dynamic and fairly demanding because the aromatic composition of foods, beverages, and perfumery products are frequently quite complex. There is often the impression that we enjoy just a single aroma or flavor sensation for a particular product when there are often actually numerous volatile compounds that combine together to impart a sensory impact. Only in rare cases can the characteristic aroma of a food product be narrowed down to one particular compound; for example, 4-hydroxy-3-methoxybenzaldehyde is given the common name of vanillin, and *trans*-3-phenyl-2-propenal is commonly known as cinnamaldehyde, owing to their association with the aromas of vanilla and cinnamon, respectively. However, the isolation of volatile compounds from natural flavor extracts has shown that, individual components frequently do not impart aromas that are reminiscent of the flavor substance [1]. Rather, a combination of particular volatile compounds is required, each at a specific concentration, for us to recognize an aroma as having a distinctly familiar odor quality.

The perceivable aromas of foodstuffs are typically the result of aroma-active compounds. This means that more attention should be focused on compounds with organoleptic significance rather than simply any identified volatile compound that may or may not contribute to a particular aroma. A good example is coffee: over 800 volatiles have currently been identified [2–4] and, among them, only about three dozen compounds have a considerable impact on the overall aroma [5,6]. It is the combination of these odor-active compounds that provides us with a characteristic coffee aroma. Therefore, accurate identification of each unique volatile compound that plays a role in contributing toward the overall aroma of a particular food is the ultimate goal of flavor research [1,7].

In order to determine which compounds have organoleptic qualities, techniques that involve olfactometry, or the sense of smell, are most preferable. Gas chromatography-olfactometry (GC-O) is typically used today to determine aroma-active compounds in complex matrices. As chromatographic separation proceeds, the effluent from the GC is split so that a portion of it goes to a physical detector, often FID (flame ionization detection) or MS (mass spectrometry), so that the signal can be recorded as a chromatogram. The rest of the effluent is directed to a "sniffing port" where human subjects use their noses to recognize individual aromas and make note of aroma descriptors at respective retention times.

It is expected that every odor that the human nose can detect can be related to a specific compound in the chromatogram. However, this is easier said than done because the chromatogram sometimes shows a flat baseline at the moment that an aroma is perceived at the sniffing port. This is because the human nose is often much more sensitive and selective than any physical detector, and a compound may not give rise to a signal if its concentration is below the detection limit of the physical detector. Generally, volatile aroma compounds are present at very low concentrations, typically on the parts-per-million (ppm) or parts-per-billion (ppb) level. Sometimes, aroma compounds are present even in concentrations as low as parts per trillion (ppt), which is often too minute to be seen as a distinguishable signal on the GC chromatogram. Frequently, an aroma-active compound can coelute with another compound that has a larger peak area. Small peaks that are related to compounds with organoleptic importance can often be concealed under larger peaks of little interest, complicating accurate identification [8,9]. Additionally, a compound may be "identifiable" using mass spectrometry or retention indices of pure standards, but it is possible that the odor quality or retention index of the "identified" compound will have no relation to the perceived aroma, which can be discouraging for the investigator. Therefore, in order to avoid such exasperating predicaments, the success of flavor characterization using GC-O techniques largely depends on the isolation and concentration techniques that are employed for the analysis of aroma-active compounds.

Although high-efficiency GC columns have been used to obtain better separation, it is impossible to resolve all the interested compounds. Preseparation of flavor compounds prior to GC-O analysis, where the chemical compounds of a complex sample are separated into classes based on their chemical or physical properties, has proven to be a helpful strategy. By this method the analysis can become greatly simplified and the presence of additional constituents can be revealed due to the reduced probability of coelution [8]. Therefore, by separating an aroma extract into different classes, the results from organoleptic analyses are often much clearer because each fraction is less complex than the overall extract. Some preseparation techniques include chemical manipulation, column-liquid chromatography, preparative high-performance liquid chromatography, and multidimensional gas chromatography. This chapter will provide an in-depth look at the application of some of these methods on various food matrices in order to provide some knowledge about the overall usefulness of using preseparation techniques to successfully analyze aroma-active compounds.

II. Chemical manipulation

A.   Classic acid and base separation

A classic way to achieve separation of chemical compounds is by manipulating the solubility properties of particular analytes [10]. Using aqueous solution and organic solvents, relatively complex mixtures can be fractionated into simpler compositions through a sequence of organic- or aqueous-phase extractions. This process depends on the inherent solubility factors of particular molecules and their preferential partitioning between phases. Acid/base fractionation is a common method of separating chemical compounds based on their differing solubilities in water and organic solvents by pH adjustments. By changing the pH of the aqueous phase, acidic and basic compounds, in addition to neutral compounds, can be successively separated from the original aroma extract so that the complexity of the chromatogram decreases with each step. This can simplify compound identification by mass spectrometry and organoleptic recognitions by GC-O.

The fractionation of acids and bases is done by manipulating the pH value based on the pKa or pKb values of the compounds; by changing the pH, the molecular form of ionizable organic analytes can be modified. Acidic compounds with a carboxyl group will be soluble in an organic phase, such as diethyl ether, when the carboxyl group is protonated at acidic pH. However, when the pH of the solution is made more basic (such as pH 8 or higher, depending on the pKa of the acid), the proton on the carboxyl group will be neutralized and the acidic molecule converts to its salt form, and thus preferentially becomes more soluble in the aqueous phase. After the separation, acidic compounds can be recovered by adjusting the aqueous phase to an acidic pH and then extracting with an organic solvent. Likewise, basic compounds (commonly amines, pyrazines, and pyridines) will be preferentially transferred into the aqueous phase at low pH, then once readjustment to a neutral pH occurs, the basic compounds will return to the organic phase.

Nonionizable neutral analytes can also be fractionated by manipulating their solubility in the aqueous phase; however, this practice is less explored. Numerous foodstuffs, especially fermented products, contain high concentrations of short-chain alcohols. Although most alcohols have very high sensory thresholds and thus contribute very little to the aroma, their high abundance in an aroma extract can interfere with chromatography and the identification of other aroma compounds. As a result, it is often desirable to remove alcohols from other neutral compounds prior to analysis. This can be done by mixing the aroma extract with the right proportion of water so that highly water-soluble compounds, such as short-chain alcohols, will separate out from other neutral compounds by partitioning into the aqueous phase. Water-soluble compounds can be recovered by saturating the aqueous phase with electrolytes, such as NaCl. This causes a salting-out effect, whereby the solubility of neutral organic solutes in the aqueous phase decreases as a result of increased electrolytic activity [11], allowing for the water-soluble fraction to be extracted with diethyl ether.

A typical fractionation scheme example is illustrated in Figure 5.1 to separate a cheese aroma extract into acidic, basic, water-soluble, and neutral fractions [12]. A Parmigiano-Reggiano cheese sample was first extracted by diethyl ether followed by high-vacuum distillation to remove the lipids. The distillate (aroma extract 1) was mixed with distilled water and adjusted to pH 10 with 10% NaOH. Sodium chloride was then added and the solution was separated into organic and aqueous phases in a separatory funnel. At pH 10, acidic solutes will transform to their salt form and partition into the aqueous phase but other organic solutes will remain in the organic phase. The diethyl ether phase was labeled (extract 2) and saved. The aqueous phase, containing the acidic compounds, was then adjusted to pH 2 with 2N H_2SO_4 and reextracted with diethyl ether, causing the acidic organic compounds to partition out of the aqueous phase. The organic portion was then dried with anhydrous $MgSO_4$ and concentrated under a nitrogen stream as the "acidic fraction."

Extract 2 was further fractionated; saturated NaCl solution was added and the pH was adjusted to 2 with 2 NH_2SO_4 so that the basic organic compounds partitioned into the aqueous phase, then the diethyl ether phase was removed and labeled (extract 3). The aqueous phase was then adjusted to pH 10 with dilute NaOH and extraction with diethyl ether yielded the "basic fraction," which was then dried over anhydrous Na_2SO_4, filtered, and concentrated. "Extract 3" was washed with water and the water washes were

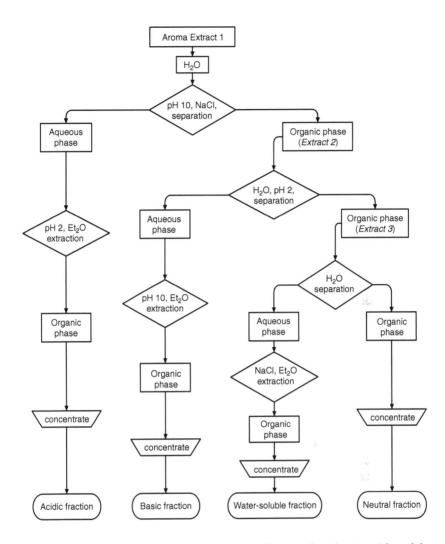

Figure 5.1 Aroma extraction and fractionation diagram for classic acid and base separation.

combined and saved, and the organic phase was dried, concentrated, and labeled as the "neutral fraction." The collected water washes from "extract 3" were extracted with diethyl ether with the addition of NaCl to decrease the aqueous solubility of the neutral water-soluble compounds so that extraction with diethyl ether would result in the "water-soluble fraction," which was then dried, filtered, and concentrated. Aroma compounds in each fraction of the Parmigiano-Reggiano sample that were identified are summarized in Table 5.1 to Table 5.3.

Straight short-chain fatty acids were the major components in the acidic fraction (Table 5.1). Acetic, butanoic, hexanoic, and octanoic acids had strong

Table 5.1 Aroma Compounds in the Acidic Fraction of Parmigiano-Reggiano Cheese

Retention Index (HP-FFAP)	Compound	Aroma character	Aroma intensity
1460	Acetic acid	Vinegar	Strong
1525	Propanoic acid	Sour	Weak
1535	2-Methylpropanoic acid	Cheesy	Very weak
1588	Butanoic acid	Cheesy, sweaty	Strong
1630	3-Methylbutanoic acid	Cheesy, sweaty	Very weak
1698	Pentanoic acid	Sour, cheesy	Very weak
1797	Hexanoic acid	Cheesy, sweaty	Strong
2008	Octanoic acid	Cheesy, sweaty	Strong
2184	Decanoic acid	Soapy, waxy	Weak
2538	Dodecanoic acid	Soapy	Very weak

aromas and most likely contributed strong sweaty, cheesy, and lipolyzed notes to the cheese. Pentanoic, heptanoic, decanoic, and dodecanoic acids, along with 2-methylpropanoic and 3-methylbutanoic acids, may also contribute to the overall aroma but were found to have much weaker intensities. Other compounds found in the acidic fraction included benzoic and phenylacetic acid but their aroma contributions to the cheese were very limited.

Table 5.2 Aroma Compounds in the Basic Fraction of Parmigiano-Reggiano Cheese

Retention index (HP-FFAP)	Compound	Aroma character	Aroma intensity
1303	2,5-Dimethylpyrazine	Baked, nutty	Very weak
1308	2,6-Dimethylpyrazine	Nutty, Chocolate	Very strong
1321	2,3-Dimethylpyrazine	Nutty, Coffee	Strong
1341	2-Ethyl-5-methylpyrazine	Baked, nutty	Very weak
1395	Trimethylpyrazine	Roasted, Baked	Strong
1408	5-Ethyl-2-methyl-pyridine	Baked, Roasted	Strong
1435	3-Ethyl-2,5-dimethylpyrazine	Roasted, Baked	Strong
1470	5-Ethyl-2,3-dimethylpyrazine	Roasted, Baked	Strong
1484	Tetramethylpyrazine	Baked	Weak
1521	2,3,5-Trimethyl-6-ethylpyrazine	Baked	Very strong
1615	2-Acetylpyridine	Baked	Weak
1609	Phenylacetaldehyde	Green, floral	Very strong
1616	2-Furanmethanol	Baked	Strong
1931	3-Methylbutaneamide	Baked	Very weak
1980	Unknown	Popcorn	Strong
2216	3-Phenylpyridine	Baked	Very weak

Table 5.3 Aroma Compounds in the Water-Soluble Fraction of Parmigiano-Reggiano Cheese

Retention index (HP-FFAP)	Compound	Aroma character	Aroma intensity
884	2-Propanol	Alcohol, fruity	Weak
924	Ethanol	Alcohol, fruity	Weak
1000	2-Butanol	Fruity	Weak
1006	1-Propanol	Fruity	Weak
1060	2-Methylpropanol	Green, Fruity	Weak
1113	2-Pentanol	Fruity	Weak
1103	1-Butanol	Green, Fruity	Weak
1169	3-Methyl-1-butanol	Green Fruity	Very weak
1274	3-Methyl-3-buten-1-ol	Fruity	Very weak
1277	3-Methyl-2-buten-1-ol	Fruity	Weak
1607	Dimethyl-dihydro-2(3H)-furanone	Baked	Strong
1616	2-Furanmethanol	Baked	Strong
1760	Unknown	Green	Weak
1788	Unknown	Baked	Strong

The basic fraction (Table 5.2) consisted of alkylpyrazines and alkylpyridines in which these compounds had characteristic nutty, baked, and roasted-like aromas. The strongest aroma compounds in this fraction were 2,6-dimethylpyrazine, 2,3-dimethylpyrazine, and 2,3,5-trimethyl-6-ethylpyrazine, along with trimethylpyrazine, 5-ethyl-2-methylpyridine, and 3-ethyl-2,5-dimethylpyrazine.

When performing fractionation in this manner, the basic and neutral fractions are often combined in order to save time during analysis. However, due to the fact that many alkylpyrazines are present at low concentrations, it is rare to obtain reliable mass spectra for these compounds when they are combined with neutral compounds. Fractionation of basic compounds into a separate group is useful in order to achieve reliable mass spectra, along with suitable GC-O recognitions, for proper identification by minimizing interference from other compounds.

For most aroma analysis, fractionation of a flavor extract into acidic, basic, and neutral fractions will provide enough separation for further GC-O analysis. For fermented products where alcohols are high, a separation of the water-soluble fraction is very useful. The most abundant compounds in the water-soluble fraction of the Parmigiano-Reggiano extract were various short-chain alcohols, and most of them contributed very weak fruity notes to the cheese (Table 5.3). Thus they can be eliminated for further extensive examination. Overall, fractionation of acidic and basic compounds, along with the removal of alcohols, greatly simplified the composition of the extract so that other important neutral compounds could be identified.

Fractionation of water-soluble compounds is particularly useful for alcoholic beverages where this approach can be effectively illustrated in the analysis of Maotai liquor (a popular Chinese spirit). The Maotai liquor was

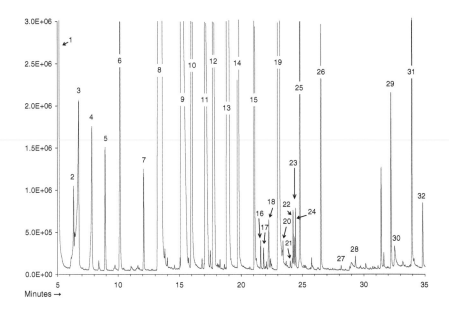

Figure 5.2 Acidic fraction of Chinese Maotai liquor. GC-MS conditions: ZB-FFAP column (30 m × 0.32 mm ID, 0.5 μm thickness). Carrier gas: helium at 2 ml/min. The oven temperature was 40°C for 2 min, increased to 230°C at a rate of 6°C/min, and maintained at 230°C for 15 min. Injector temperature was 250°C. MS 70 eV with scan range of 35 to 300.
Key: (1) ethanol; (2) 2-butanol; (3) 1-propanol; (4) 2-methylpropanol; (5) 1-butanol; (6) 3-methylbutanol; (7) 3-hydroxy-2-butanone; (8) ethyl 2-hydroxypropanoate; (9) acetic acid; (10) furfural; (11) propanoic acid; (12) 2-methylpropanoic acid; (13) butanoic acid; (14) 3-methylbutanoic acid; (15) pentanoic acid; (16) 2-methylpentanoic acid; (17) 2-butenoic acid; (18) 4-methylpentanoic acid; (19) hexanoic acid; (20) 2-methyl-hexanoic acid; (21) 5-methylhexanoic acid; (22) phenylethanol; (23) 2-methyl-2-pentenoic acid; (24) 2,4-dimethyl-2-pentenoic acid; (25) heptanoic acid; (26) octanoic acid; (27) nonanoic acid; (28) decanoic acid; (29) benzoic acid; (30) furan-2-carboxylic acid; (31) phenylacetic acid; and (32) phenylpropanoic acid.

first diluted with water to give an alcohol level of 14% then the aroma compounds were extracted with Freon 11, followed by acidic and basic fractionations. As shown in Figure 5.2, Figure 5.3, and Figure 5.4, the acidic fraction was comprised mainly of acidic compounds, the basic fraction consisted mostly of pyrazines, and the water-soluble fraction was primarily alcohols. Because alcohols and acids were the most abundant compounds in the overall aroma extract, the removal of these compounds made the identification of other neutral aroma compounds feasible (Figure 5.5).

   Fractionation using classic acid and base methods can also be performed to obtain specific types of compounds. For example, by adjusting the pH level within a more narrow range, it is possible to partially separate carboxy acids from phenols. When the aqueous phase is adjusted to pH 8, most carboxy acids will turn to the salt forms and become soluble in the aqueous

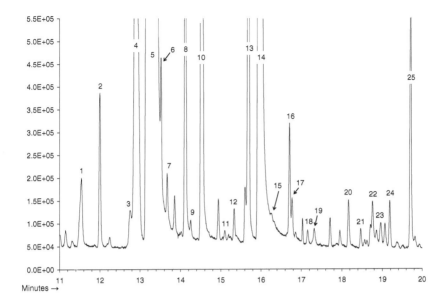

Figure 5.3 Basic fraction of Chinese Maotai liquor (GC-MS conditions were the same as Figure 5.2).
Key: (1) 2-methylpyrazine; (2) 3-hydroxy-2-butanone; (3) 2,5-dimethylpyrazine; (4) 2,6-dimethylpyrazine; (5) ethyl 2-hydroxypropanoate; (6) 2,3-dimethylpyrazine; (7) 2-hydroxy-3-pentanone; (8) 2-ethyl-6-methylpyrazine; (9) 2-ethyl-5-methylpyrazine; (10) 2,4,5-trimethylpyrazine; (11) 2,6-diethylpyrazine; (12) 2,5-dimethyl-3-ethylpyrazine; (13) 2,3-dimethyl-5-ethylpyrazine; (14) furfural; (15) 2,3,5,6-tetramethylpyrazine; (16) 2,3,5-trimethyl-6-ethylpyrazine; (17) 2-acetylfuran; (18) benzaldehyde; (19) ethyl 2-hydroxyhexanoate; (20) 5-methyl-2-furfural; 21) 2-butyl-3,5-methylpyrazine; (22) 2-acetylpyridine; (23) 2-acetyl-5-methylfuran; (24) 2-acetyl-6-methylpyridine; and (25) 2-furanmethanol.

phase, while the phenols will stay in the organic phase. When the pH is further raised to pH 10 or above, the phenols become ionized and are soluble in water, thus facilitating the analysis of this particular group of compounds (Figure 5.6 and Figure 5.7).

Similarly, fractionation can be achieved by chemical reactions to separate compounds based on the functional group. Since the volatile composition of many food matrices contain carbonyl compounds, it is possible to selectively remove methyl ketones and aldehydes from the system. For analytes in the aqueous phase, the addition of sodium bisulfite will form water-soluble bisulfite-carbonyl complexes so that they will not be extracted into the organic phase; similarly, if the analytes of interest are contained within the organic phase, the addition of a 2,4-dinitrophenylhydrazine solution will cause water-soluble adduct formations with carbonyl-containing compounds [13,14]. In both cases, the carbonyl-containing compounds will not be extracted into the organic phase, producing a carbonyl-free extract for analysis, if desired.

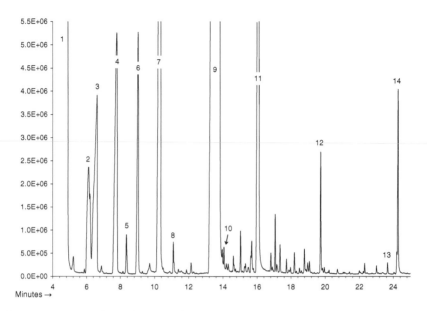

Figure 5.4 Water-soluble fraction of Chinese Maotai liquor (GC-MS conditions were the same as Figure 5.2).
Key: (1) ethanol; (2) 2-butanol; (3) 1-propanol; (4) 2-methylpropanol; (5) 2-pentanol; (6) 1-butanol; (7) 3-methylbutanol; (8) 1-pentanol; (9) ethyl 2-hydroxypropanoate; (10) 3-ethoxy-1-propanol; (11) furfural; (12) 2-furanmethanol; (13) benzenemethanol; and (14) phenylethanol.

The fractionation procedures can easily be done in a separatory funnel. However, if the volume of the aroma extract is small, a specialized Mixxor™ extractor can be used [8,15]. The general operation of the Mixxor™ apparatus involves placing a small volume of an aroma extract (for example, the aqueous condensate obtained from steam distillation or the organic distillate from high-vacuum distillation) into the reservoir of the extractor. Either an immiscible organic solvent or water is then added, depending on the nature of the extract (organic solvent for aqueous extract, water for organic extract). The system is extracted by moving the piston up and down several times. With this extraction device, there is no need for lengthy vigorous shaking to achieve efficient separation; it is estimated that six piston movements are approximately equivalent to 40 shakes with a separatory funnel [16]. Once phase separation occurs within the reservoir, the solvent can be forced into the narrow axial channel where it can be easily removed with a syringe for analysis.

The steps for fractionation can be performed sequentially [14] in the Mixxor™ by first adjusting to alkaline pH followed by extraction with diethyl ether. A small portion of the organic phase, which will not contain acids, can be removed for GC analysis, then the solution in the Mixxor™ is made acidic and reextracted with the same diethyl ether already in the system. An aliquot of this organic phase, which does not contain bases, is removed for analysis. If desired, the carbonyl compounds can also be

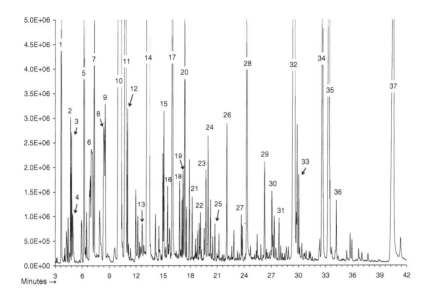

Figure 5.5 Neutral fraction of Chinese Maotai liquor (GC-MS conditions were the same as Figure 5.2).
Key: (1) 1,1-diethoxyethane; (2) ethyl propanoate; (3) ethyl 2-methylpropanoate; (4) 1,1-diethoxy-2-methylpropane; (5) ethyl butanoate; (6) 1,1-ethoxy-3-methylbutane; (7) 2-methylpropanol; (8) ethyl pentanoate; (9) 1-butanol; (10) 3-methylbutanol; (11) ethyl hexanoate; (12) 1-pentanol; (13) ethyl heptanoate; (14) ethyl 2-hydroxypropanoate; (15) ethyl octanoate; (16) 1-heptanol; (17) furfural; (18) 2-acetylfuran; (19) benzaldehyde; (20) ethyl 2-hydroxyhexanoate; (21) 5-methyl-2-furfural; (22) ethyl decanoate; (23) 2-furanmethanol; (24) diethyl butanedioate; (25) 1,1-diethoxy-2-phenylethane; (26) ethyl phenylacetate; (27) benzenemethanol; (28) phenylethanol; (29) ethyl tetradecanoate; (30) 4-methylphenol; (31) ethyl pentadecanoate; (32) ethyl hexadecanoate; (33) ethyl 2-hydroxy-3-phenylpropanoate; (34) ethyl Z-9-octadecanoate; (35) ethyl Z,Z-9,12-octadecadienoate; (36) ethyl Z,Z,Z-9,12,15-octadecatrienoate; and (37) hexadecanoic acid.

removed by returning the pH to neutral then adding sodium bisulfite or 2,4-dinitrophenylhydrazine. An example of the application of this method was given by Parliment for the analysis of roasted coffee [8]. The aroma of roasted coffee is comprised of a very complex mixture of volatile compounds that are present across a broad range of concentrations [17]. Therefore, the preseparation of volatile compounds into less complicated fractions is a necessity in order to successfully analyze and identify the aroma-active compounds in this matrix.

Roasted coffee beans were ground and steam distilled with water using a modified Likens–Nickerson simultaneous distillation–extraction apparatus [18,19] with pentane:diethyl ether (1:1) as the solvent. The Mixxor™ extractor was then used to successively remove classes of compounds from the aroma extract. Owing to the complexity of the sample, the aqueous phase was removed from the system at each step and fresh reagents were added so that

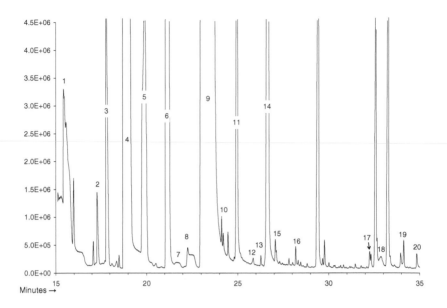

Figure 5.6 Acidic/phenolic fraction from Chinese Yanghe Daqu liquor obtained by extracting with Freon 11 and then adjusting to pH 8 (GC-MS conditions were the same as Figure 5.2).
Key: (1) acetic acid; (2) propanoic acid; (3) 2-methylpropanoic acid; (4) butanoic acid; (5) 3-methylbutanoic acid; (6) pentanoic acid; (7) 2-methylpentanoic acid; (8) 4-methylpentanoic acid; (9) hexanoic acid; (10) 5-methylhexanoic acid; (11) heptanoic acid; (12) phenol; (13) 4-ethyl-2-methoxyphenol; (14) octanoic acid; (15) 4-methylphenol; (16) nonanoic acid; (17) benzoic acid; (18) Z,Z-9,12-octadecadienoic acid; (19) phenylacetic acid; and (20) phenylpropanoic acid.

only the original organic phase was further extracted. (Note: If the investigator is interested in studying the analytes that are removed, then the aqueous portion from each step can be reextracted after making the proper pH adjustments.) By removing the acidic and basic fractions from the original coffee extract, which included compounds such as 2-methylbutanoic acid, 3-methylbutanoic acid, and some ethyl- and methyl-substituted pyrazines, it was possible to reveal a volatile compound that contributes considerably to the overall aroma of coffee: 2-furfurylthiol. Many sulfur compounds are quite important to the aroma of roasted coffee [17] and this particular sulfur compound, 2-furfurylthiol, has been found to be a key odorant by way of sensory omission (triangle) experiments [20]. Therefore, the application of this preseparation technique was advantageous by allowing for better resolution of the chromatographic peak for a significant aroma contributor.

In general, chemical manipulation methods are relatively rapid and easy to employ, require minimal equipment, and have reproducible results. Furthermore, sensory analyses can be conducted on each individual fraction (i.e., acid, basic, neutral, water-soluble, etc.) to evaluate the compounds that

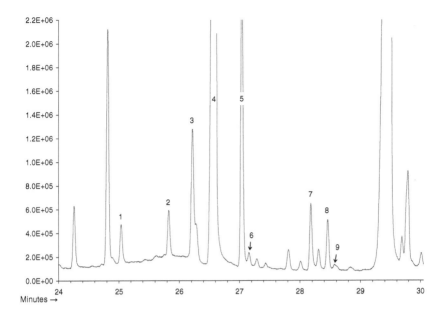

Figure 5.7 Phenolic fraction from Chinese Yanghe Daqu liquor obtained by extracting with Freon 11 and then adjusting to pH 10 (GC-MS conditions were the same as Figure 5.2).
Key: (1) 2-methoxy-4-methylphenol; (2) phenol; (3) 4-ethyl-2-methoxyphenol; (4) octanoic acid; (5) 4-methylphenol; (6) 3-methylphenol; (7) nonanoic acid; (8) 4-ethylphenol; and (9) 3-ethylphenol.

are most important to the overall aroma. However, because many of these separation methods are achieved via pH adjustments of the aqueous phase, it is possible that artifacts could be created. Under acidic conditions, esters are susceptible to hydrolysis reactions, producing the corresponding alcohols and acids. Likewise, acetal derivatives of aldehydes and ketones can be generated through dehydration reactions in the presence of alcohols or diols. Similarly, hydrolysis of aldehydes and ketones can be either acid- or base-catalyzed, creating geminal diols [21]. Nevertheless, artifact formation is a process that can occur during any preparative method and is something that must be kept in mind when performing aroma analysis.

III.   Column-liquid chromatography

The development of column-liquid chromatography has contributed tremendously to flavor analysis. Preparative column chromatography has been widely used in flavor studies as a valuable tool to isolate volatiles at very low levels or to separate interested compounds from a complex matrix. Although the previous section discussed separation techniques using liquid–liquid partitioning, column chromatography involves the isolation of analytes using liquid–solid partitioning. Problems that are associated with liquid–liquid extractions, such as incomplete separations and disposal of large amounts of organic solvents, can be avoided by use of a solid stationary phase to extract target analytes from a solution [22]. Analytes are first extracted from a complex matrix, then particular classes of compounds can be isolated from the extract owing to interactions with the functional groups of a solid stationary phase. In this section, three major applications of preparative column-liquid chromatography are discussed, including silica gel normal-phase column chromatography, reversed-phase column chromatography and ion-exchange chromatography.

A. Normal-phase chromatography

1. Silica gel column chromatography

Silica gel, also known as silicic acid or silica, is the most widely used chromatographic adsorbent for separating organic compounds (molecular weight up to 2000 daltons [23]) of varying polarity via conventional normal-phase column chromatography. The behavior of silica gel as a chromatographic adsorbent depends on the size of the silica particles, its hydroxyl group distribution, and the shape and size of the pores within the three-dimensional siloxane structure [23,24]. The silica hydroxyls, or silanols, are primarily responsible for retaining analytes within the column in which the solute molecules are attracted to the polar surface of the silica gel through dipole–dipole interactions and hydrogen bonding [23,25,26]. In order to use the silica gel column as an adsorbent, it must first be "activated" by removing water. This is necessary because the capacity factor of silica gel decreases with increasing water content; water molecules lower the surface activity due to their strong interaction with the polar silica surface [23]. To prepare a silica gel column for the separation of aroma compounds, a slurry of silicic acid is poured into a glass column. The column is then conditioned by first rinsing with a polar solvent, such as diethyl ether, methanol, or dichloromethane, to remove any contaminants. Then a dry nonpolar solvent, such as pentane or hexane, is rinsed through to remove the polar solvent so that all surface hydroxyls become available for adsorptive interactions with the solutes of the aroma extract [23].

Once the extract is applied to the column, fractionation into separate classes can be achieved due to polarity differences between different types of aroma compounds [22]. The separation process relies on the inherent affinity of the solute compounds to either remain adsorbed onto the stationary phase or to be eluted with the mobile phase. The first fraction is obtained by eluting with a pure nonpolar solvent, then successive elutions are performed by stepwise increases of the solvent polarity with the final elution done with a pure polar solvent. For example, start with pentane, then use pentane:diethyl ether (90:10), followed by pentane:diethyl ether (80:20), etc., until the last elution with 100% diethyl ether (this could also be followed by a final rinse with an even more polar solvent, such as methanol). By successively passing different solvent mixtures through the column, the adsorbed compounds can be successfully separated into different fractions based on their interactions with the column and the solvents [22]. When the chromatographic conditions (such as the amount of silica gel to be used and its activity level, the solvent types and their specific ratios, along with elution rates, column height and diameter, and so on) are carefully selected, a complex aroma mixture can be well separated [27]. Using normal-phase chromatography for class separation, the eluting order of classes is usually as follows: hydrocarbons, esters, aldehydes and ketones, alcohols, lactones, and, finally, acids.

A technique for class separation of flavor volatiles by silica gel column chromatography was first developed by Murray and Stanley in 1968 [28]. A Teflon tube with a glass wool plug was packed with partially deactivated silica gel. A green pea aroma extract was added to the top of the column, followed by more silica gel and a second glass wool plug. The column was then inverted into a beaker containing the developing solvent. When development was complete, the Teflon tube was cut into small sections and then the adsorbed volatiles from each section were eluted with ether and concentrated. By using this method of fractionation, GC-MS analysis revealed that alcohols and hydrocarbons were well resolved from other volatiles, although aldehydes, ketones, and esters had some overlap between fractions. Although this method was quite tedious, as the tube needed to be cut in order to fractionate the analytes, it did provide some valuable information.

This fractionation strategy for identifying complex flavor mixtures has been applied in the analysis of a wide variety of foods, including cheese [12], wine [29–32], fruits [33–36], coffee and teas [37–39], and distilled liquor [40–42]. For example, silica gel was used to further separate the neutral fraction of a Parmigiano-Reggiano cheese aroma extract to achieve better identification of the neutral compounds [12]. A mini silica gel column (250 mg) was conditioned, then the neutral fraction was applied to the column and a pentane-diethyl ether gradient was used to achieve further fractionation (Table 5.4). Esters and one sulfur compound were recovered in the combined pentane, pentane:diethyl ether (90:10), and pentane:diethyl ether (70:30) fractions. Aldehydes and ketones were mainly found in the pentane:diethyl ether (50:50) fraction, while phenylacetaldehyde and *p*-ethylbenzylacetaldehyde were identified in a more polar fraction (pentane:diethyl ether, 30:70). After the column was washed with 100% diethyl ether, recovery of the most polar compounds in the neutral fraction (2-furanmethanol, γ-octalactone, benzothiazole and δ-decalactone) was achieved.

The benefits of class fractionation with silica gel can also be seen in the following example [43], where a concentrated "Marionberry" blackberry extract was prepared in dichloromethane from 1 kg of fruit. Because GC-O-MS identification of the aroma-active compounds in the original extract was not successful due to numerous overlaps of peaks, fractionation was necessary in order to reduce the complexity of the extract. After removing the acidic fraction, the neutral fraction was further separated using a silica gel column (8 g, 15 cm height, 12 mm ID) which was washed with methanol and equilibrated

Table 5.4 Aroma Compounds Identified in Neutral Fraction of Parmigiano-
Reggiano Cheese Using Normal Phase Chromatography and GC-Olfactometry

Retention index (HP-FFAP)	Compounds	Aroma character	Aroma intensity
Pentane, Pentane:Diethyl Ether (90:10) and Pentane:Diethyl Ether (70:30) Fractions			
955	Ethyl propanoate	Fruity	Strong
1022	Ethyl butanoate	Fruity	Very strong
1224	Ethyl hexanoate	Fruity	Very strong
1422	Ethyl octanoate	Fruity	Strong
1615	Ethyl decanoate	Fruity	Strong
1675	Ethyl 9-decenoate	Fruity	Strong
1820	Ethyl dodecanoate	Fruity	Weak
1377	Dimethyltrisulfide	Cabbage	Very strong
Pentane:Diethyl Ether (50:50) Fraction			
864	2-Methylbutanal	Cocoa	Strong
912	3-Methylbutanal	Malty	Strong
1073	2-Methyl-2-butenal	Malty	Strong
1160	2-Heptanone	Blue cheese-like	Very strong
1268	2-Octanone	Blue cheese-like	Weak
1202	(*E*)-2-Hexenal	Green	Weak
1372	2-Nonanone	Fruity	Weak
1458	Methional	Potato	Very strong
1368	2,4-Hexadienal	Green	Strong
1509	(*E*)-2-Nonenal	Green, floral	Weak
1579	2-Undecanone	Green, baked	Strong
1793	2-Tridecanone	Fatty	Very weak
Pentane:Diethyl Ether (30:70) Fraction			
1609	Phenylacetaldehyde	Green	Very strong
1949	*p*-Ethyl benzylacetaldehyde	Baked green	Strong
Diethyl Ether Fraction			
1616	2-Furanmethanol	Baked	Strong
1902	Benzothiazole	Rubbery	Weak
1990	δ-Decalactone	Coconut	Strong

in pentane. The extract was added and successive fractions were obtained by washing the column with 40 ml of each of the following solvents at a flow rate of 1 ml/min: pentane (fraction 1); pentane:diethyl ether, 95:5 (fraction 2); pentane:diethyl ether, 90:10 (fraction 3); pentane:diethyl ether, 50:50 (fraction 4); and diethyl ether (fraction 5). After fractionation, identification was greatly simplified due to decreased peak crowding.

As shown by the chromatogram in Figure 5.8, the pentane fraction contained only hydrocarbon compounds; although many fruits often contain high amounts of hydrocarbons, this class does not generally contribute to the aroma of the fruits. Hence, the pentane fraction can effectively remove odorless hydrocarbon compounds so that other odor-active compounds can be more easily identified. Compounds in fraction 2 were primarily esters with a trace amount of hexanal and nonanal (Figure 5.9). A small amount of these two aldehydes in fraction 2 is reasonable because they are found in high concentrations in Marionberries. Compared with many other fruits, the ester content of Marionberry is relative low; therefore, the fractionation of

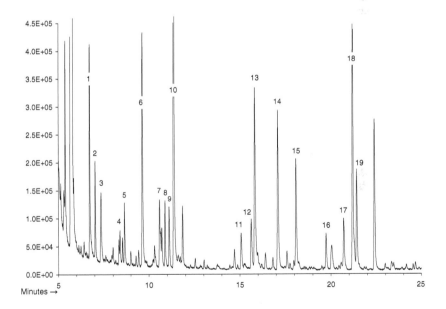

Figure 5.8 Chromatogram of fraction 1 (pentane) obtained by normal-phase chromatography of neutral fraction of Marionberry extract. GC-MS conditions: ZB-Wax column (30 m × 0.25 mm ID, 0.25 μm thickness). Carrier gas: helium at a constant flow rate of 2 ml/min. The oven temperature was 40°C for 2 min, increased to 230°C at a rate of 6°C/min, and maintained at 230°C for 15 min. Injector temperature was 250°C. MS 70 eV with scan range 35 to 300.
Key: (1) α-pipene; (2) toluene; (3) camphene; (4) ethylbenzene; (5) xylene; (6) limonene; (7) γ-terpinene; (8) β-ocimene; (9) p-cymene; (10) α-terpinolene; (11) α-cubebene; (12) α-ylangene; (13) α-copaene; (14) isoledene; (15) β-caryophyllene; (16) calarene; (17) α-elemene; (18) α-farnesene; and (19) δ-cadinene.

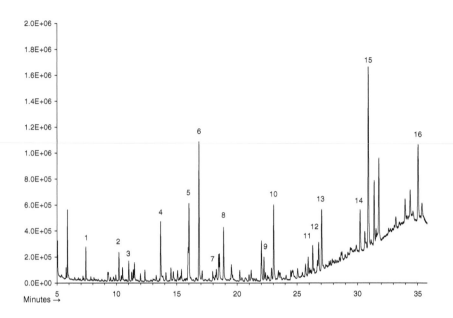

Figure 5.9 Chromatogram of fraction 2 (pentane:diethyl ether, 95:5) obtained by normal-phase chromatography of neutral fraction of Marionberry extract (GC-MS conditions were the same as Figure 5.8).
Key: (1) hexanal; (2) ethyl hexanoate; (3) hexyl acetate; (4) nonanal; (5) theaspirane B; (6) theaspirane A; (7) methyl decanoate; (8) ethyl decanoate; (9) methyl dodecanoate; (10) ethyl dodecanoate; (11) methyl tetradecanoate; (12) isopropyl tetradecanoate; (13) ethyl tetradecanoate; (14) methyl hexadecanoate; (15) ethyl hexadecanoate; and (16) ethyl Z-9-octadecenoate.

esters makes it possible to identify the odor-active esters present at trace levels, such as ethyl hexanoate and isoamyl isovalerate. Ketones and aldehydes were successfully fractionated into fraction 3 (Figure 5.10); some odor-active compounds identified in this fraction included 2-heptanone, *trans*-2-hexenal, 1-octen-3-one, nonanal, and benzaldehyde. The fractions eluted with pentane:diethyl ether (50:50) and 100% diethyl ether, fraction 4 and fraction 5, respectively, both had the majority of hydroxyl-containing compounds (Figure 5.11 and Figure 5.12). Many of the alcohols in these two fractions were determined to be odor-active through GC-O, including 2-heptanol, *trans*-2-hexenol, 1-decanol, α-terpineol, and linalool, along with phenylethanol and 4-phenyl-2-butanol. Depending on the composition of the aroma extract and the objective of fractionation, elution with pentane:diethyl ether (50:50) could be omitted since this fraction may yield similar compounds to those obtained with 100% diethyl ether.

Overall, fractionation of the Marionberry extract by normal-phase chromatography allowed for easier GC-O recognitions and subsequent mass spectral identifications. For example, the full extract gave a mixed aroma at approximately 17.9 min with poor chromatographic peak shape due to multiple

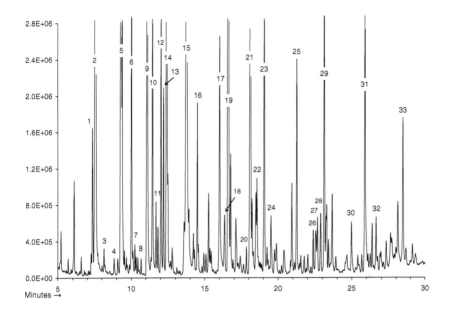

Figure 5.10 Chromatogram of fraction 3 (pentane:diethyl ether, 90:10) obtained by normal-phase chromatography of neutral fraction of Marionberry extract (GC-MS conditions were the same as Figure 5.8).
Key: (1) butyl acetate; (2) hexanal; (3) 2-methylbutyl acetate; (4) 2-ethyl-2-butenal; (5) 2-heptanone; (6) *trans*-2-hexenal; (7) ethyl hexanoate; (8) *cis*-4-heptanal; (9) hexyl acetate; (10) octanal; (11) 1-octen-3-one; (12) *cis*-3-hexenyl acetate; (13) 2-heptenal; (14) *trans*-2-hexenyl acetate; (15) nonanal; (16) 2-octenal; (17) decanal; 18) L-camphor; (19) benzaldehyde; (20) *trans*-2-*cis*-6-nonadienal; (21) 2-undecanone; (22) 1-*p*-men-then-9-al; (23) *trans*-2-decenal; (24) 2-butyl-2-octenal; (25) undecenal; (26) 2-tride-canone; (27) *trans*-damascenone; (28) dihydroxy-ionone; (29) 4-hydroxy-ionone; (30) β-ionone; (31) caryophyllene oxide; (32) salvial-4-(14)-en-1-one; and (33) 6,10,14-tri-methyl-2-pentadecanone.

coeluting compounds. However, after fractionation, several odor-active com-pounds in that region became clear due to class separation of the respon-sible aroma compounds. These included β-caryophyllene (wood, spice) in fraction 1, methyl decanoate (fruity, sweet) in fraction 2, 2-undecanone (floral, citrus) in fraction 3, and mesifurane (sweet, fruity, caramel) in fraction 5. Another interesting compound identified after fractionation was 4-(*p*-hydroxyphenyl)-2-butanone (raspberry ketone), which has a high boiling point (200°C) and elutes quite late from the GC column (retention index on ZB-Wax = 2985). It was not identified during GC-O of the original extract, possibly because when the aroma profile is very complex, even a well-trained panelist will tire toward the end of the run and become less sensitive to odorants. However, fractionation enabled the successful recognition of this compound with its characteristic raspberry aroma.

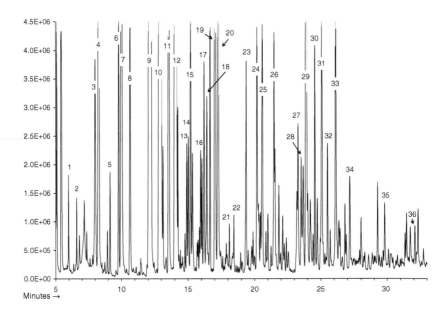

Figure 5.11 Chromatogram of fraction 4 (pentane:diethyl ether, 50:50) obtained by normal-phase chromatography of neutral fraction of Marionberry extract (GC-MS conditions were the same as Figure 5.8).
Key: (1) 2-pentanone; (2) ethyl acetate; (3) 3-pentanol; (4) 2-pentanol; (5) 3-penten-2-ol; (6) 2-methylbutanol; (7) *trans*-2-hexenal; (8) 1-pentanol; (9) 2-heptanol; (10) 1-hexanol; (11) *cis*-3-hexenol; (12) *trans*-2-hexenol; (13) 1-octen-3-ol; (14) heptanol; (15) 6-methyl-5-hepten-2-ol; (16) *trans,trans*-2,4-heptadienal; (17) 3-ethyl-4-methylpentanol; (18) 2-nonanol; (19) linalool; (20) 1-octanol; (21) *cis*-non-3-en-2-ol; (22) *trans*-2-decenol; (23) nonanol; (24) α-terpineol; (25) 2-undecanol; (26) 1-decanol; (27) *p*-cymen-8-ol; (28) benzylacetone; (29) benzyl alcohol; (30) phenylethanol; (31) β-ionone; (32) benzothiazole; (33) 4-phenyl-2-butanol; (34) cinnamaldehyde; (35) torreyol; and (36) cinnamyl alcohol.

While the method of silica-gel fractionation has proven effective for the separation of aroma compounds based on class properties, this technique also has some inherent limitations. Although silica can be used with practically all solvents, the presence of water will decrease its activity, and some solutes or solvents can remain fixed within the column [26], resulting in irreversible adsorption. Additionally, it is possible for silica to induce artifact formation by catalyzing the degradation of labile compounds [44]. López et al. [45] first proposed the use of LiChrolut-EN, composed of poly(styrene-divinylbenzene), or PS-DVB, as an alternative packing material for normal-phase chromatography in order to study the volatile aroma compounds in a wine matrix. PS-DVB, which is a polymeric porous resin that is now available in commercially prepared SPE (solid phase extraction) cartridges, has some advantages over silica resins: first, it is stable over the entire pH range (1 to 14); second, it can more effectively retain nonpolar organic compounds; and third, its hydrogen-bonding capacities are less and, therefore, irreversible adsorption can usually be avoided [22,46].

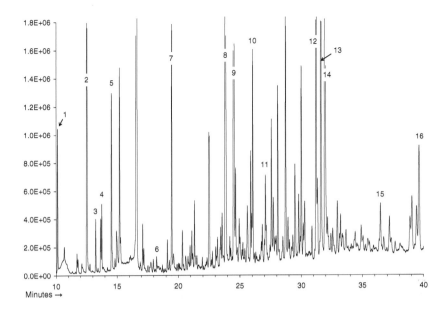

Figure 5.12 Chromatogram of fraction 5 (diethyl ether) obtained by normal-phase chromatography of neutral fraction of Marionberry extract (GC-MS conditions were the same as Figure 5.8).
Key: (1) 1-pentanol; (2) 1-hexanol; (3) *cis*-3-hexenol; (4) *trans*-2-hexenol; (5) *cis*-linalool oxide; (6) mesifurane; (7) 2,3-epoxyhexanol; (8) benzyl alcohol; (9) phenylethanol; (10) 4-phenyl-2-butanol; (11) hydrocinnamyl alcohol; (12) 8-acetoxylinalool; (13) cinnamyl alcohol; (14) 1-hydroxylinalool; (15) 3-hydroxy-damascone; and (16) δ-dodecalactone.

Culleré et al. [44] optimized the use of PS-DVB for SPE fractionation of a wine extract. After conditioning a LiChrolut-EN column with methanol and dichloromethane, a small aliquot of aroma extract was carefully introduced. The aroma fractions were obtained by sequentially washing the column with pentane, pentane: dichloromethane (9:1), and dichloromethane, then the three fractions were concentrated and analyzed by GC-MS. Ethyl esters and some other nonpolar compounds were recovered in the first fraction. In the second fraction, alcohols, volatile phenols, and some other compounds of intermediate polarity were recovered, while fatty acids and vanillin derivatives were concentrated in the third fraction. The recovery of the volatiles in each fraction was nearly 100%, with better reproducibility than normal-phase silica columns. Additionally, the maximum amount of extract that can be successfully retained by LiChrolut-EN columns is about eight times higher than that of silica owing to its larger surface area [44]. Although the selectivity for the adsorption of analytes on LiChrolut-EN resin can be more specific than silica gel [45], this sorbent reduces the possibility of artifact formation that results from the increased activity of silica gel.

2. *Reversed-phase chromatography*

In addition to normal-phase-liquid chromatography, silica can also be modified for use in reversed-phase liquid chromatography. Generally, **reversed-phase** chromatography uses a relatively nonpolar stationary phase with a polar mobile phase. In order to create a nonpolar surface on silica, aliphatic hydrocarbon chains, commonly n-octyldecyl (C_{18}) or n-octyl (C_8) chains, are chemically bonded to the surface silanols [47]. However, there will almost always be some hydroxyl groups that are left unmodified but the presence of the residual silanols has been reported to assist in the separation of certain polar analytes [22]. Fractionation is performed by starting with solvents that have high polarity, such as water, acetone, or methanol, then decreasing the solvent polarity stepwise so that the final elution is done with a nonpolar solvent. The interactions between the solute and the bonded-reversed phase are very similar to that of the normal phase; however, because the surface is much more hydrophobic, reversed-phase columns will retain analytes in the opposite order so that the solutes are eluted in order of decreasing polarity [23,48]. Therefore, when using reversed-phase resins, the elution order for solutes of similar size is alcohols and amines, acids, esters, aldehydes and ketones, then hydrocarbons [49]. Of course, within each class, molecules that have greater chain lengths (i.e., more carbon atoms) or bulky ring structures will generally have greater retention times on the column.

The choice of using either normal or reversed-phase columns depends on the properties of the analytes to be separated, where one phase will usually give better results than the other. One advantage of using reversed-phase-liquid chromatography is that the separation of aqueous solutions is possible because the stationary phase will not strongly bind with water, as seen with normal-phase silica resin. Nevertheless, it must be remembered that most aroma compounds are not very soluble in highly aqueous mobile phases. Additionally, the use of reversed-phase resins is still limited because the presence of water in the fractions indicates that an additional extraction step is necessary before performing GC analyses.

B. *Ion-exchange chromatography*

Ion-exchange chromatography is another method widely used in food analysis in which charged substances are separated via interactions with column materials that carry an opposite charge. There are several types of ion-exchange phases available, the choice of which depends on the properties of the target analytes which must be capable of having an ionic charge when in aqueous or organic solutions. Unlike the use of normal-and reversed-phase columns for aroma analysis, in which the primary objective is to fractionate an extract into different classes, the goal of ion exchange is to effectively isolate one particular type of compound from the extract. The general procedure is to load a solution onto the ion-exchange column, wash away undesired components, and then elute the desired analytes with another solvent for collection and subsequent analysis [50]. The primary mechanism for the retention of the analyte on the ion-exchange stationary phase is through electrostatic attractions between the analyte and the bound functional group (ligand) attached to the column material [51].

Ion-exchange ligands can be either negatively or positively charged. Negatively charged ion exchangers adsorb cations and positively charged ones adsorb anions; hence they are named cation exchangers and anion exchangers, respectively. The exchange process can be dependent on the pKa of the ligand, which further categorizes ion-exchange chromatography as strong or weak. For strong ion-exchange columns, the functional group remains charged over a wide pH range. For weak ion exchange, the charge of the ligand is greatly dependent on the pH of the solution and is therefore is charged over a limited pH range. (Note: Strong or weak in this sense does not refer to the binding strength provided but rather to the ion-exchange ligand as a strong or weak acid or base [52].) An oppositely charged counterion is attached to the bound ligand so that, when the solution containing ionic analytes is passed over the stationary phase, the charged analyte will exchange places with the counterion and thus become trapped by the resin [53]. Impurities, such as neutral compounds or anything with an incorrect charge for electrostatic attraction to the stationary phase, can be easily washed from the column. Finally, in order for the target analytes to be released for collection, a solution is added to the column that has a pH that will neutralize either the target compound or the bonded functional group, effectively disrupting the electrostatic force that held them together.

In flavor analysis, this technique is frequently applied to separate fatty acids from complex matrices, such as dairy products. Ion-exchange column chromatography is a suitable method to quantitatively isolate free fatty acids (FFA) because these molecules can be trapped by the column through electrostatic processes. In early work, silicic-acid-KOH columns were used to isolate FFA from cheese and butter [54–57]. Because the pH of the column was made alkaline by the presence of KOH, the FFA were captured as potassium salts. After washing the column to remove the neutral lipids, the FFA were eluted with acidified diethyl ether. Although this method can isolate FFA, the preparation of both the sample and the silicic-acid-KOH column can be tedious, and prolonged fat contact with the KOH column can greatly increase lipid hydrolysis, causing variability of the FFA profile.

Aluminum oxide, or alumina, is another widely used adsorbent and is similar to silica due to its numerous active surface hydroxyls [47]. Because alumina can also function as an ion exchanger [26], this stationary phase has been used to extract FFA from triglyceride mixtures [58,59]. Although it is possible for the high surface activity of common alumina to induce hydrolysis of triglycerides, partial deactivation with 5 to10% water prior to use will avoid any hydrolytic reactions. When the sample extract is applied to the column, FFA are adsorbed to the alumina so that undesired neutral lipids can be removed with a solvent wash. However, in order to desorb the FFA for analysis, it is necessary to dry the alumina prior to extraction with acidified diethyl ether. With this method, recovery of FFA from cheese samples was found to be 95% where the loss occurred mainly with butanoic acid [58].

Strong anion-exchange resins, such as Amberlyst A-26 and Dowex 1 × 2, have also been used in conjunction with methylation processes to aid in the isolation of FFA [60–64]. After triglycerides are washed from the column, the absorbed FFA are methylated by mixing the dried resin with HCl-methanol creating fatty acid methyl esters, or FAMEs. These resins have also been used as heterogeneous catalysts where methylation of FFA is carried out [62]. Following methylation, FAMEs are then extracted with solvent and can be directly analyzed by GC. By using the ion-exchange resin as a catalyst, complete methylation of FFA occurs without potential hydrolysis of triglycerides or transesterification with other lipids [65]. However, the reproducibility of butanoic acid is still poor owing to potential evaporation loss of methyl butanoate during the methylation steps.

For successful FFA analysis, hydrolysis of triglycerides should be avoided at all costs to prevent the overestimation of FFA concentrations. Also, since butanoic acid is an important aroma-active compound in foods such as aged cheese [66–69], this compound needs to be fully recovered in order to know the true impact it may have on the overall aroma and flavor. Therefore, as an improved alternative to the above methods, the amino-weak anion-exchange column has been employed in numerous studies for determination of FFA content [58,70–74]. This method is preferable since it does not require drying, nor is it necessary to create FAMEs prior to GC analysis. Amino ion-exchange columns are commercially available as prepacked cartridges: the stationary phase is held in place by porous frits within the small plastic column that ends in a Luer tip to facilitate transfer to a collection vial [50].

Using an amino ion-exchange column, Qian and Reineccius [70] studied the FFA profile in Parmigiano-Reggiano cheese. The lipid extract, which contained both FFA and undesired triglycerides, in hexane-isopropanol (1:1) was passed through an aminopropyl SPE column, preconditioned with heptane. After the solution had passed through, the neutral lipids were rinsed from the column with chloroform-isopropanol (2:1), and then FFA were eluted with diethyl ether containing 2% formic acid. The FFA extract was directly injected onto a GC installed with an FFAP capillary column. This method is simple and quick with good reproducibility where hydrolysis of triglycerides was not observed and, by using the aminopropyl stationary phase, nearly 100% recovery can be achieved for all FFA [58]. This technique has been used to analyze and compare the fatty acid composition of several different types of cheese [75], as shown in Table 5.5.

Since conventional acidic fractionation by pH adjustment cannot entirely remove all acids, especially long-chain fatty acids, the application of the ion-exchange technique has also been attempted to completely remove acids from the neutral and basic fraction [76]. Concentrated wine extract, prepared in diethyl ether-pentane, was loaded onto an amino ion-exchange column. Washing the column with diethyl ether yielded the neutral and basic compounds and, by rinsing with diethyl ether containing 1% formic acid, the acidic fraction was recovered. By applying ion-exchange column chromatography, the wine aroma extract was easily separated into two fractions. A total of 87 aroma-active compounds were revealed (Table 5.6a and Table 5.6b), in which 11 compounds were found in the acidic fraction and 76 compounds in the neutral and basic fraction.

Table 5.5 FFA and OAV Determined in Various Cheeses Using Amino Exchange Chromatography and GC Analysis

Cheese type (Brand/Origin)	Butanoic acid (C4:0)		Hexanoic acid (C6:0)		Octanoic acid (C8:0)		Decanoic acid (C10:0)	
	(ppm)	(OAV[a])	(ppm)	(OAV)	(ppm)	(OAV)	(ppm)	(OAV)
Cheddar								
Black Diamond	97	36	36	4	25	1	98	4
Bongards	21	8	9	1	11	0.6	52	2
Cracker Barrel	62	23	21	2	15	0.8	65	3
Kraft	173	64	52	6	21	1	79	4
New York	98	36	41	4	26	1	89	4
Vermont	189	70	59	6	26	1	101	4
West Coast	39	14	16	2	11	0.6	43	2
Wisconsin	60	22	16	2	14	0.7	56	2
Feta (traditional)	1567	580	677	74	221	12	410	19
Pecorino Romano	651	241	424	46	213	11	391	18
Provolone	512	190	243	26	62	3	164	7
Roquefort	434	162	334	36	373	20	1313	60
Parmigiano-Reggiano	397	147	249	27	146	8	374	17

[a] Odor activity value; equal to concentration divided by odor threshold.

C.   Isolation of thiols by ion-exchange chromatography

Although advances in analytical techniques and instrumentation have led to great progress in flavor analysis, the analysis of sulfur compounds still remains challenging. Sulfur compounds have very low odor thresholds; therefore, very low concentrations can still contribute to aroma. Since degradation

Table 5.6a Important Aroma-Active Compounds in the "Acidic Fraction" of 2000 Vintage Pinot Noir Wine

Compound	Aroma descriptor	Intensity[a]
3-Methyl butanoic acid	Rancid, cheesy, sweaty	12
Hexanoic acid	Sour, acid	12
Acetic acid	Acidic, vinegar	9
2-Methyl propanoic acid	Woody, sour	7
Benzeneacetic acid	Musty, fruity	7
Butanoic acid	Cheesy, sweaty	6
2-Methyl butanoic acid	Sweaty	6
Octanoic acid	Goaty	5
t-2-Hexenoic acid	Acidic, green	4
Benzoic acid	Pungent	3
Propanoic acid	Sour	2

[a] Fifteen-point intensity scale; value determined through GC-O analysis.

Table 5.6b Some Aroma-Active Compounds in the "Basic and Neutral Fraction" of 2000 Vintage Pinot Noir Wine

Compound	Aroma descriptor	Intensity[a]	Compound	Aroma descriptor	Intensity
Isoamylalcohol	Unpleasant, musty, nail polish	14	Trans-geraniol	Green, floral	5
2-Phenyl-ethanol	Rose	14	Propanol	Floral, tulip	5
Damascenone	Green, fruity	12	Diethyl malate	Green, fruity, caramel	5
Methionol	Potato, vegetable	10	3-Hydroxy-4-phenyl-2-butanone	Floral, fruity, berry	5
Ethyl hexanoate	Fermenta-tion, sweet, fruity	10	Ethyl-2-hydroxy-3-phenyl-propanoate	Black pepper	5
Ethyl phenyl-acetate	Floral, pungent, sweet	10	Ethyl dihydrocin-namate	Fruity sweet	5
Isoamyl acetate	Banana	9	Benzothiazole	Dry grass	5
Isobutyl alcohol	Nail polish	9	5-Methylthia zole	Smoky earthy	5
Benzyl alcohol	Floral, leaf	9	Ethyl benzoate	Floral, tea	5
Nonalacone	Floral, coconut	9	Di-isobutyl succinate	Fruity	4
Diethyl azeleate	Smoky, floral	8	Guaiacol	Chemical pungent, sweet	4
Ethyl 2-hydroxy-hexanoate	Floral, jasmine	8	Phenol	Floral, sweet	4
Ethyl isobutyrate	Floral, fruity, sweet	8	Ethyl cinnamate	Floral, cherry, fruity	4
Ethyl 3-phenylpro-pionate	Floral,	8	1-Ethoxy-1-pen-toxy-ethane	Metal, burning	4
Ethyl isovalerate	Sweet, grass	7	α-Terpinene	Green, lemon	4
Phenethyl acetate	Wine, floral	7	Isoamyl hexanoate	Fruity, green	3
1-Hexanol	Vegetable, fruity	7	Benzyl acetate	Floral, herb	3
cis-3-Hexenol	Green	6	Tetradecanal	Fatty, oily	3
Ethyl 2-hydroxy isovalerate	Fatty, milky, sweet	6	β-ionone	Floral, woody	3

Table 5.6b (*Continued*) Some Aroma-Active Compounds in the "Basic and Neutral Fraction" of 2000 Vintage Pinot Noir Wine

Compound	Aroma descriptor	Intensity[a]	Compound	Aroma descriptor	Intensity
Ethyl-2-hydro-xybutyrate	Floral	6	1-Octen-3-ol	Mushroom, earthy	3
Vanillin	Vanilla powder	6	3-Ethyl-pent-anol	Floral, ester	3
Ethyl anthranilate	Sweet fruity	6	Fenchyl alcohol	Fresh grass, floral	3
Ethyl butyrate	Floral, berry	6	Methyl anthranilate	Fruity	3
Ethyl 2-methylbuty-rate	Fruity, orange, sweet	5	Trans-3-hexenol	Green	3
Ethyl octanoate	Green, fruity	5	Ethyl-3-hydroxyoc-tanoate	Smoky, stimulate	3
α-Terpinolene	Sweet, nutty	5	Benzothiazole	Woody, stimulate	3
Ethyl 3-methyl thiopropionate	Smoky	5	Geranic acid	Fruity	3
Ethyl caprate	Fruity, grape	5	2-Octanol	Earthy, aromatic	2
Ethyl lactate	Cooked, fruity	5	Citronellol	Green, fruity	2
Trans-linalool oxide	Wet, woody, grass	5	Isoeugenol	Roasted, woody	2

[a] Fifteen-point intensity scale; value determined through GC-O analysis.

can easily occur during sample preparation and analysis procedures, a novel approach toward the extraction of volatile thiols from wine was proposed by Tominaga et al. in 1998 [77]. This method involved the combination of chemical manipulation for extraction of thiols from the matrix through the formation of reversible complexes and a strongly basic anion-exchange column as a clean-up procedure for isolation of the thiols.

Wine samples were extracted with dichloromethane, and the organic phase was reextracted with *p*-hydroxymercuribenzoic acid (sodium salt solution) to trap the thiols [77,78]. The aqueous phase, containing the thiol complexes, was then brought to pH 7 with dilute HCl and applied to a strongly basic anion-exchange column (Dowex 1X2 chloride form) where any *p*-hydroxymercuribenzoate, either "free" or thiol-complexed, became attached to the resin. The column was rinsed to remove impurities and the thiols were released from *p*-hydroxymercuribenzoate by washing the column with a cysteine solution. The thiol-containing eluate was reextracted with dichloromethane to remove excess cysteine, which is not soluble in organic solvent.

Analysis by GC-MS identified 4-mercapto-4-methylpentan-2-one and also 3-mercaptohexan-1-ol and 3-mercaptohexyl acetate as contributors to the complex aroma of Sauvignon blanc wine [77].

Even though a handful of other compounds are also extracted into the aqueous phase of *p*-hydroxymercuribenzoate [79], ion exchange allows those impurities to be removed for the effective isolation of volatile thiols from a wine matrix. However, it has been suggested that pH 7 was not sufficient to maintain all *p*-hydroxymercuribenzoic acid in its water-soluble salt form [80] as it is important that the mercury-trapping salt will quantitatively partition into the aqueous phase during thiol extraction and also that proper ion exchange will occur with the strongly basic column. Nevertheless, by using internal standard, this method allowed for the analysis of low levels of sulfur compounds at ppb levels in wines. Due to their low concentrations, it is highly likely that without the application of this preseparation technique, other compounds would have masked the chromatographic peaks of these aroma-active volatile sulfur compounds.

In summary, column fractionation eliminates the exposure of aroma compounds to strong alkaline and acidic conditions, thus reducing artifact formation and improving quantitative analyses. Overall, the use of column chromatography can effectively separate compounds based on their functional groups, such as hydrocarbons, esters, aldehydes and ketones, and so on. Today, there is an increasing availability of different types of stationary phases, allowing excellent separations for practically all types of trace analytes [22]. In general, column-liquid chromatography is a useful method of separating aroma extracts into less complex fractions so that positive GC-O or MS identifications are more clear-cut. Although the methods for preseparation of aroma extracts mentioned thus far have proven extremely useful, technological advances have allowed the establishment of preseparation techniques with even greater resolution of aroma-active compounds.

IV.   Preparative and multidimensional HPLC and GC

For the analysis of aroma compounds and successful detection, there is a need for efficient enrichment of the analytes, which are generally present at ppb levels. Additionally, in order to reduce peak overlap, separation must occur in more than one dimension [81]. Therefore, the techniques discussed in this section can assist in solving the problems associated with coeluting compounds. The method of conventional column-liquid chromatography is relatively easy, although sometimes time consuming, but because it does not require sophisticated equipment and expensive adsorbents, it is a commonly used procedure for fractionation of aroma extracts. However, more technologically advanced methods have been developed that employ HPLC and GC techniques for preseparation of compounds prior to analytical determinations which can eliminate the need for tedious classical chemical fractionation in many cases. These separation methods can be performed both off-line and online and are generally accepted to be more efficient than other separation techniques, often with better resolution for trace analytes. Combinations of preseparation techniques, such as conventional silica gel column fractionation followed by further fractionation by either HPLC or GC, are often used as a means to thoroughly separate an aroma extract prior to analysis. As with all methods of fractionation, the goal is to separate any sensorially important compounds into several portions so that identification, including mass spectral and GC-O analyses, is simplified due to the decreased overlap of chromatographic peaks.

A.   Preparative HPLC and GC techniques

Preparative chromatography using HPLC and GC prior to analytical separations has been used in recent decades as a means of off-line isolation and concentration of particular compounds in an aroma extract. In order to utilize these instruments as techniques for preseparation, some preliminary method development must be performed to determine the optimum conditions for fractionation, such as solvent composition, flow rate, and the best suited stationary phase. Peaks of interest should have sufficient baseline separation so that the corresponding eluates can be contained within a single fraction. Furthermore, if specific analytes are to be isolated, then their respective retention times will need to be known for proper collection.

Preparative HPLC separation of complex mixtures is performed based on an array of different physical properties where both normal-phase and reversed-phase chromatography are widely used as preseparation methods. It is reported that reversed-phase resins have a higher loading capacity than untreated silica resin; therefore, larger samples can be loaded onto these columns which could be quite useful for preparative HPLC [82]. Another advantage of using HPLC is that it uses nondestructive methods of detection, generally ultraviolet light, and therefore the entire injected sample can be recovered. A sequential separation strategy for complex aroma extracts using HPLC was first suggested by Teitelbaum in 1977 [9]. A crude aroma extract of cocoa butter was cleaned and fractionated on a conventional silica gel column, then the fractions with "good cocoa butter aroma" were pooled for fine separation by HPLC with multiple columns. An adsorption column (Partisil 10) was used, primarily because it is able to handle a wide range of classes; the fractions that emerged with the solvent front were further separated by a reversed-phase column and the later eluting fractions were applied to a normal-phase column for further separation. However, determining which column is more effective for separation depends entirely upon the composition of the extract, which is why preliminary work is necessary for successful fractionation by HPLC.

The analysis of alcoholic beverages is often difficult due to the presence of numerous volatile constituents formed during fermentation. These compounds can mask the important aroma-active compounds during chromatographic analyses. Therefore, Ferreira et al. (1999) thoroughly investigated the usefulness of preparative reversed-phase HPLC for fractionating wine extracts with regard to separation efficiency, as affected by solvent composition and maximum injection volume, as well as the possibility of artifact formation during the chromatographic process [83]. In order to avoid the use of toxic solvents to allow for auxiliary sensory analysis, a water-ethanol gradient was tested and found to be sufficient for eluting all compounds from a Kromisil-C_{18} column. Their solvent gradient scheme was found to be advantageous since all major aroma compounds eluted in the first part of the chromatogram and were well separated, allowing for relatively large sample volumes (between 2 to 4 ml of extract) to be injected without serious peak distortion. Any chemical changes induced by the column, as related to the perceived aroma of each wine fraction, were found to be of minor importance. Although it was determined that the developed method was simple and highly robust for aroma fractionation, the authors acknowledged that the major drawback of using reversed-phase chromatography is that fractions must be reextracted prior to GC-MS analysis.

In a study on oxidation-related odorants in wine, fractionation was achieved using both normal- and reversed-phase HPLC [84]. The preparation of the aroma extract involved the removal of fatty acids and fusel alcohols since these compounds can cause column saturation during chromatography. The extract was then injected separately onto two HPLC columns. Nonpolar fractions were obtained on a normal-phase column with a mobile phase of diethyl ether:dichloromethane:hexane (8:2:90), where collections were taken at regular intervals, concentrated, and analyzed by GC-O-MS. A reversed-phase column was used to separate the polar fraction using a water-ethanol gradient as the mobile phase, where each polar fraction acquired from this preparative HPLC technique was further extracted by dichloromethane prior to concentration and subsequent GC-O-MS analysis. The polar fractions were found to contain compounds such as methional, sotolon, and furfural, and the nonpolar fractions contained eugenol, benzaldehyde, and 1-octen-3-ol, all of which were determined to be partly responsible for the oxidative aroma of wine.

Sensory panels are typically used to evaluate HPLC fractionation so that the proper fractions can be selected for further GC-MS analysis. This approach has been used to investigate the impact odorants of red wine aroma [85]. A trained sensory panel evaluated the odor quality of individual HPLC fractions where only the odor-active fractions were selected for further analysis by GC-O-MS. In another study dealing with the identification of compounds in strawberry jam extract, a comparison was made before and after HPLC fractionation using a diol-bonded column with a pentane and diethyl ether gradient [86]. Approximately 150 compounds could be positively identified by GC-MS in the sample that was preseparated by HPLC, in contrast to only 60 compounds before fractionation was performed. Based on identification results, aroma recombination can be conducted to compare imitation aroma mixtures with the original fractions.

While preseparation using HPLC is a useful approach to help solve some difficult problems in aroma analysis, it also has disadvantages that can limit its application. During fraction collection, the eluting solutions are often in direct contact with the atmosphere, so it is possible for evaporation to occur whereby some highly volatile analytes could be lost [87]. Also, the analysis may require repeated injections in order to obtain sufficient concentrations in each fraction for successful identification by GC-MS. Another drawback of this technique is that when reversed-phase resins are used for fractionation, there is often the need to reextract the aroma volatiles from the fractions due to the presence of water or alcohol, which are incompatible with GC analyses.

In the past, preparative work using GC has been less common than HPLC due to its low sample load and difficulty in efficiently collecting or condensing the eluates from the gas stream [88]. Additionally, high temperatures are normally required to vaporize analytes for GC analysis, which could potentially introduce artifacts into fractions obtained in this manner. Thus, during analytical chromatography of these fractions, it would be impossible to know if all identified compounds were originally present in the extract or if some were the result of the preseparation technique. Furthermore, because GC detection methods are generally destructive, the only way to know when particular compounds are eluting and still be able to collect them is to split the column near the end so that a small portion goes to the detector.

Nevertheless, several studies have been successfully performed using GC as an off-line method of fractionation [89–93]. The types of columns that are generally used for preparative GC include packed stainless steel or wide-bore capillary columns because large injection volumes are required to obtain sufficient concentrations within each fraction. Although a wide selection of stationary phases are available for GC columns, separation is mainly due to differing vapor pressures. As a result, when using GC columns of similar polarity but differing stationary phases, only small variations will be seen in resolution and order of appearance of analytes. Therefore, the column chosen for preparative separation is typically of opposite polarity to the column that is used for analytical separation. In some applications involving preparative GC, fraction collection is performed by trapping the GC effluent of a specified retention window into a volume of solvent [90,91,93]. Improved techniques involve the use of a cold trap (approximately –50°C, glass) to condense the analytes from particular portions of the effluent, then rinsing the trap with solvent later to recover the analytes [89,94]. After trapping, an aliquot of the solvent is injected onto an analytical GC column for further separation. In recent years, superior enhancements of preparative GC have been developed using automated systems, as discussed in the next section.

B.   *Multidimensional gas chromatography*

Multidimensional gas chromatography (MDGC), involving online separation and enrichment, greatly increases the separation ability of gas chromatography [95] and has been available to chemists for decades. In the early stages, coupled-column GC analyses were achieved by connecting two columns together using intermediate pressure-tuning and flow-modulated operation.

By using two different columns, each of opposite polarity, an overall improvement in the degree of separation of a complex mixture can be accomplished [96]. However, this tandem operation of GC columns does not actually represent multidimensionality, but rather resembles the use of a mixed-stationary-phase column [95]. True MDGC involves a process known as orthogonal separation where a sample is first dispersed in time by one column, then simplified subsamples are applied onto another column for further separation on a secondary time axis that is completely independent of the first. For accurate identification, the data from both retention-time axes are needed because they provide complimentary information [97], where identification by multidimensional techniques can be more reliable than with single-column separation because two independent retention times are obtained for each analyte [98].

Today, modern MDGC techniques can be generally divided into two classes: (1) conventional or "heart-cut" MDGC and (2) comprehensive two-dimensional gas chromatography (GC × GC). Several reviews have been published that provide extensive details about the principles involved when using MDGC to successfully analyze complex samples [95,97,99–101]. While multidimensional HPLC techniques, both heart-cut and comprehensive, are possible, they are not generally used for aroma analysis but rather in studies dealing with larger-molecular-weight compounds, such as proteins [98].

1. Conventional two-dimensional GC

Conventional two-dimensional GC is achieved by using coupled capillary columns where a small portion, or heart-cut, of the effluent from the first (preseparation) column is transferred to the second (analytical) column. The concept of conventional MDGC is almost identical to that of preparative GC operations, as described earlier, where one column is used to obtain a partially separated fraction of a complex aroma mixture, which is then reinjected onto another GC column, usually with an opposite stationary phase, for further separation. The only difference is that with MDGC there are no requirements for manual collection of the effluent obtained from the preseparation column because the two columns are directly connected. In order to effectively analyze a sample by conventional MDGC, a preliminary injection must be performed with only the preseparation column in order to obtain an uncut chromatogram so that the desired times for heart-cutting can be determined.

With proper system design, selected cuts from the preseparation column will be transferred to the main analytical column with minimal peak distortion. Cryotrapping of the volatiles after elution from the preseparation column is the most effective way to ensure that all analytes within the heart-cut enter the analytical column at the same time. By simultaneously revolatilizing the analytes from the cold trap, the separation that occurs in the second column becomes orthogonal to the first so that the two columns do not simply perform as a mixed-stationary-phase column [102]. In fact, cryotrapping between columns can be quite advantageous for dilute solutions because solvents with low boiling points can be partly or totally flushed from the trap prior to release of the analytes, allowing for online enrichment [103].

An illustration of such an application can be seen in a study on the fungus-type aroma found in wine grapes contaminated by powdery mildew [104]. MDGC was performed by using a multicolumn switching system (Gerstel, Germany) coupled with mass spectrometry (MDGC-MS). Preseparation was conducted on a nonpolar capillary column (HP-5) and a polar column (BP-20) was used for analytical separation. A 3.5-min heart-cut was made from the first column at a predetermined retention time, and the fraction was cryotrapped at –50°C at the head of the second column. Once the entire heart-cut was transferred, the trap was immediately warmed and analytical separations using a ramped temperature program began. By utilizing MDGC, the presence of 1-octen-3-one (strong mushroom odor) was revealed in the heart-cut chromatogram and, through AEDA, this compound was determined to be among the most potent volatile odorants of the mildewed grapes.

Because the second column in the MDGC system is only injected with a small portion of the total sample at one time, a large quantity of the sample can be injected onto the first column without the worry of the chromatographic band smearing during analytical separations [105]. Therefore, trace compounds can be easily enriched for more successful detection and identification. The limits of sensitivity for MDGC, as compared to single-column GC, were tested during the analysis of a peppermint oil sample using Fourier transform infrared spectroscopy (FTIR) as the method of detection [105]. In the MDGC system, an RSL-200 column was used as the preparative column where heart-cuts were sent to a DB-Wax analytical column; the data was compared to that obtained by a DB-5 single-column GC system. With the use of conventional MDGC-FTIR, the factor of apparent sensitivity enhancement was 5 to 27 times greater than compared to single-column GC-FTIR, further supporting the benefits of using MDGC for analyses of trace aroma compounds.

By performing enough heart-cuts, it is possible to analyze the entire sample by MDGC as shown by Gordon et al. for a flue-cured tobacco essential oil sample [106]. MDGC was conducted on a multidimensional capillary-switching device (Scientific Glass Engineering, Australia) with a coupled Carbowax precolumn and an OV-1701 analytical column. A total of 23 heart-cuts were employed across the entire chromatogram of the preseparation column, each with a duration of a few minutes. Over 300 volatile compounds were well separated, where 20% were identified for the first time in flue-cured tobacco. In some cases, isomer identification was made possible because of the increased separation efficiency by MDGC. Even though this approach of preseparation was quite effective, the analysis of all 23 heart-cuts was time consuming and required a total of 48 h of instrument time.

The study of enantiomers can be a challenging area for flavor chemists, but a GC column with the proper stationary phase can allow for successful separations, especially when MDGC is utilized. The extent of interaction between a particular enantiomer and the stationary phase depends on the degree of stearic hindrance or how close the compound can approach the stationary phase. The use of MDGC for entantiomeric separations has been described [102], where the transfer of narrow heart-cuts from a nonselective preseparation column to an analytical column with an enantioselective stationary phase allowed for good separation of targeted chiral compounds. This separation is successful because, when the stationary phase is also chiral by nature, one enantiomer in the sample is likely to interact closely with the stationary-phase surface, whereas others will be stearically excluded and thus have decreased retention within the column [107]. A similar approach using conventional MDGC-MS was successfully utilized for separating four different naturally-occurring enantiomers of theaspirane found in several types of fruit and tea leaves [108].

2.   Comprehensive two-dimensional GC

Comprehensive MDGC is among the most powerful two-dimensional gas chromatographic techniques today. Unlike conventional MDGC where only particular segments are transferred from the preseparation column onto the analytical column, comprehensive MDGC, or GC × GC, involves the transfer of the entire effluent from the first column onto a second orthogonal column by way of a modulation interface so that complete two-dimensional data can be obtained for the entire run of the first column. The operation of the modulator involves the generation of narrow injection bands from the first column, which are continuously and individually sent to the secondary column for final separation. GC × GC requires that the second column can operate quickly enough to generate a complete set of data during the time that a single peak elutes from the first GC column, generally within 5 sec [95,97]. The data from both time axes are combined to create a set of coordinates for each peak so that the resultant chromatogram is actually a two-dimensional (2D) plane rather than a straight line (Figure 5.13). Peak-area information can be obtained by summing the integration over both dimensions.

By using data from two dimensions, quantification can be more accurate due to a reduced chance of coeluting peaks and greater sensitivity as a result of fast-GC with the second column [100]. These features provide distinctive advantages for GC × GC for achieving complete qualitative and quantitative analyses of aroma extracts. Although gas chromatography is used in both dimensions, the mechanisms of separation within the two columns are different and independent of each other. The first column separates compounds based mainly on differences in volatility. However, when the effluent is collected in the modulator, the analytes that comprise each injection band have practically identical volatilities. Therefore, when the bands are pulsed onto the second column, separation will occur predominantly on the basis of polarity differences or hydrogen bonding with the stationary phase [99,100]. As the first column run proceeds, the temperature program used for the second column can be modified to account for decreasing analyte volatility, so that the effect of volatility on the separation mechanism of the second column can be eliminated and the process can still be deemed as truly orthogonal [100].

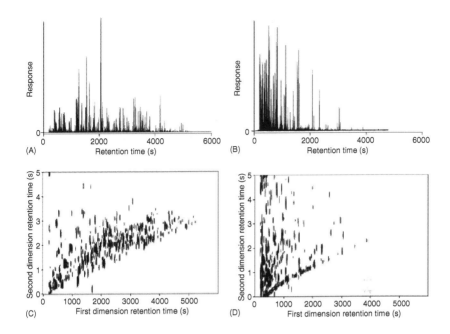

Figure 5.13 Total ion chromatograms and their respective two-dimensional contour plots for an Arabica coffee extract separated by GC × GC using two different column sets: polar × nonpolar (A and C) along with nonpolar × polar (B and D). (Reprinted from Ryan, D., Shellie, R., Tranchida, P., Casilli, A., Mondello, L., and Marriott, P., Analysis of roasted coffee bean volatiles by using comprehensive two-dimensional gas chromatography-time-of-flight mass spectrometry, *J Chromatogr A*, Vol. 1054, 57–65, 2004. With permission.)

In comprehensive GC × GC, the two columns perform independently of each other; therefore, the overall peak capacity becomes the product of the capacities for each column. Sensitivity is enhanced as a result of zone compression as the effluent from the first column travels through the modulator and enters the second column in a narrow band, producing sharper signals when the peak reaches the detector [95]. Because analytes elute from the second column so quickly, data acquisition must be fast enough for proper detection. Although FID has been commonly used, it is not very useful for identification purposes, which is why GC × GC coupled with an MS detector is often desired for aroma analysis. Time-of-flight mass spectrometry (TOF-MS) and rapid-scanning quadrupole mass spectrometry (qMS) have both been used as effective detection methods for GC × GC in order to obtain mass spectral information [109,110]. Only a few reports for the quantification of aroma compounds by GC × GC have been published to date [111,112]. However, in recent years, several studies have applied comprehensive GC × GC for qualitative determinations of aroma extracts.

Dimandja et al. [113] analyzed the volatile components of peppermint and spearmint essential oils using both one-dimensional GC and comprehensive

GC × GC in order to compare the resolution abilities of the two techniques. GC × GC allowed for the identification of 89 and 68 peaks in the peppermint and spearmint oil samples, respectively, compared to only 30 and 28 peaks by way of one-dimensional GC. Additionally, GC × GC was reported to deliver approximately six resolved peaks per min; but for the same sample using ordinary GC with a similar stationary-phase column, the delivery speed is only about one resolved peak per min [113]. The 2D chromatogram that is obtained through GC × GC is essentially a "fingerprint" for that particular sample, and comparisons can be made between fingerprints of different samples to easily investigate constituent similarities, as was done for these mint-oil samples where it was determined that there were 52 compounds in common. Headspace solid-phase microextraction (HS-SPME) combined with comprehensive GC × GC has also been examined for suitability and compatibility in order to sample, separate, and analyze the volatile compounds associated with garlic flavor [114]. By comparing the results to one-dimensional GC, this approach demonstrated a 10- to 50-fold increase in sensitivity, up to 10 times higher peak capacity, and superior separation power, all of which allow for more in-depth aroma analysis.

In the analysis of lavender essential oil, many more components were identified within the two-dimensional separation space, compared with one-dimensional GC where identification was achieved by converting the planar chromatogram back into linear form to obtain information about retention indices [109]. This essential oil contained several monoterpenes and sesquiterpenes, in which their oxygenated derivatives tended to elute very closely from the first column. However, because these compounds have a wide polarity range, the secondary separation in GC × GC allowed for polarity differences to play a large role and effectively spread these compounds across a wider region of the two dimentional chromatogram. Enantiomeric analyses have also been successfully performed using comprehensive GC × GC for the study of monoterpenes in *Melaleuca alternifolia* essential oil, whereby the first column contained an enantioselective stationary phase [115]. It should be noted that (entantio-GC) × GC is less demanding than GC × (enantio-GC) due to the requirements of fast elution on the secondary column along with sufficient resolution of all target enantiomer pairs. However, future developments of enantioselective fast-GC columns may allow for wider use of GC × (enantio-GC) [95]. Some other studies using GC × GC for efficient separation of various aroma extracts include the analysis of roasted coffee [116,117], dairy products [118,119], and strawberries [120].

Although the instrumentation can be quite expensive, the use of comprehensive GC × GC for volatile aroma analysis has exponentially increased over the past few years as methodologies have become more established and systems have become commercially available. The only drawback of GC × GC for aroma analysis is that the separation proceeds too quickly in the second column to allow for simultaneous GC-O analysis. The human nose simply cannot recognize individual odors at the rate that compounds elute from the second column. However, performing GC-O analysis is an essential tool in the proper identification of aroma compounds. Therefore, if MDGC and GC-O are to be performed at the same time, conventional heart-cut MDGC would be better suited because the elution rate from the second column is more reasonable, even though it is more time consuming when heart-cutting the entire first column chromatogram. Overall, the application of MDGC, both conventional and comprehensive, has allowed the advanced separation of complex aromas by using state-of-the-art instrumentation.

V.   Summary and outlook

In conclusion, there are numerous techniques available that can be used to simplify the analysis of complex aroma mixtures. Regardless of which method is used, successful aroma analysis demands that no components of a complex food matrix are lost or artificially created during any part of the extraction procedure, chromatographic fractionation, or analytical separation in order to ensure accurate reports of the volatile compounds that are responsible for providing characteristic aromas. Classic acid and base fractionation allows for the separation of the analytes based on pH adjustments, and although this method may require large amounts of organic solvents, there is no need for expensive equipment or materials to achieve beneficial simplification of the aroma extract. The use of silica gel with normal phase column-liquid chromatography allows for effective class fractionation to occur based on functional group interactions between the analytes, the column, and the mobile phase. Ion-exchange chromatography can allow specific analytes to be targeted for removal from a solution based on electrostatic interactions where a variety of stationary phases are available for a wide range of applications.

While HPLC and GC are commonly used as analytical instruments, they can also be utilized off-line to achieve fractionation of complex mixtures, where HPLC has a greater applicability in aroma analysis because of its ease of use compared to off-line GC-fractionation methods. However, when online preseparation is desired, multidimensional GC is fast becoming the tool of choice due to its superior ability to resolve volatile compounds. MDGC still has some limitations with regard to simultaneous olfactometric determinations, but with continuously improving technologies, the use of MDGC may soon fulfill all the requirements necessary for successful aroma analysis. Overall, the techniques presented in this chapter allow for the simplification of complicated identification procedures. Regardless of the exact method utilized, preseparation of aroma extracts is always advantageous in order to successfully analyze the chemical compounds responsible for providing us with particular sensory characteristics.

References

1. McGorrin, R.J., Character impact compounds, in *Flavor, Fragrance, and Odor Analysis*, Marsili, R.T, Ed., New York: Marcel Dekker, 2002, pp. 375–413.
2. Nijssen, L.M., Visscher, C.A., Maarse, H., and Willemsens, L.C., Eds., *Volatile Compounds in Food: Qualitative and Quantitative Data*, 7th ed., Zeist, The Netherlands: TNO Nutrition and Food Research Institute, 1996.

3. Flament, I., Coffee, cocoa, and tea, in *Volatile Compounds in Food and Beverages*, Maarse, H., Ed., New York: Marcel Dekker, 1991, pp. 617–669.
4. Flament, I., *Coffee Flavor Chemistry*, Chichester: John Wiley & Sons, 2002.
5. Kumazawa, K. and Masuda, H., Investigation of the change in the flavor of a coffee drink during heat processing, *J Agric Food Chem*, Vol. 51, 2674–2678, 2003.
6. Grosch, W., Flavour of coffee: a review, *Nahrung*, Vol. 42, 344–350, 1998.
7. Parliment, T.H. and Scarpellino, R., Organoleptic techniques in chromatographic food flavor analysis, *J Agric Food Chem*, Vol. 25, 97–99, 1977.
8. Parliment, T.H., Applications of a microextraction class separation technique to the analysis of complex flavor mixtures, in *Flavor Analysis Developments in Isolation and Characterization*, Mussinan C.J. and Morello M.J., Eds., Washington, D.C.: American Chemical Society, 1998, pp. 8–21.
9. Teitelbaum, C.L., A new strategy for the analysis of complex flavors, *J Agric Food Chem*, Vol. 25, 466–470, 1977.
10. Vogel, A., *Practical Organic Chemistry*, 3rd ed., New York: John Wiley & Sons, 1956.
11. Peng, J. and Wan, A., Effect of ionic strength on Henry's constants of volatile organic compounds, *Chemosphere*, Vol. 36, 2731–2740, 1998.
12. Qian, M. and Reineccius, G., Identification of aroma compounds in Parmigiano Reggiano cheese by gas chromatography/olfactometry, *J Dairy Sci*, Vol. 86, 1362–1369, 2002.
13. Shriner, R., Fuson, R., Curtin, D., and Morrill, T., *The Systematic Identification of Organic Compounds*, 6th ed., New York: John Wiley & Sons, 1980.
14. Parliment, T.H., Solvent extraction and distillation techniques, in *Techniques for Analyzing Food Aroma*, Marsili R.T., Ed., New York: Marcel Dekker, 1997, pp. 1–26.
15. Parliment, T.H., Solvent extraction and distillation techniques, in *Flavor, Fragrance, and Odor Analysis*, Marsili R.T., Ed., New York: Marcel Dekker, 2002, pp. 1–23.
16. Parliment, T.H., A new technique for GLC sample preparation using a novel extraction device, *Perf Flav*, Vol. 11, 1, 1986.
17. Grosch, W., Chemistry III: volatile compounds, in *Coffee: Recent Developments*, Clarke, R. and Vitzthum, O., Eds., Oxford: Blackwell Science, 2001.
18. Likens, S.T. and Nickerson, G.B., Detection of certain hop oil constituents in brewing products, *Proc Am Soc Brew Chem*, 5–13, 1964.
19. Schultz, T.H., Flath, R.A., Mon, T.R., Eggling, S.B., and Teranishi, R., Isolation of volatile components from a model system, *J Agric Food Chem*, Vol. 25, 446–449, 1977.
20. Czerny, M., Mayer, F., and Grosch, W., Sensory study on the character impact odorants of roasted Arabica coffee, *J Agric Food Chem*, Vol. 47, 695–699, 1999.
21. Carey, F.A., *Organic Chemistry*, 3rd ed., New York: McGraw-Hill, 1996.
22. Hennion, M.-C., Solid-phase extraction: method development, sorbents, and coupling with liquid chromatography, *J Chromatogr A*, Vol. 856, 3–54, 1999.
23. Unger, K.K., *Porous Silica: Its Properties and Use as Support in Column Liquid Chromatography*, Series: Journal of Chromatography Library, Vol. 16, Amsterdam, The Netherlands: Elsevier Scientific Publishing Company, 1979.
24. Wren, J.J., Chromatography of lipids on silicic acid, *J Chromatogr*, 1960, Vol. 4, 173–195.
25. Notovny, M., Chromatography on silylated surfaces, in *Bonded Stationary Phases in Chromatography*, Grushka, E., Ed., Ann Arbor, MI: Ann Arbor Science, 1974.

26. Gritter, R.J., Bobbitt, J.M., and Schwarting, A.E., *Introduction to Chromatography*, Oakland, CA: Holden-Day, 1985.

27. Scott, R.P.W., *Liquid Chromatography Column Theory*, Series: Separation Science Series, Ed., Scott R.P.W. and Simpson, C.F., Chichester: John Wiley & Sons, 1992.

28. Murray, K.E. and Stanley, G., Class separation of flavour volatiles by liquid chromatography on silica gel at 1°C, *J Chromatogr*, Vol. 34, 174–179, 1968.

29. Kotseridis, Y. and Baumes, R., Identification of impact odorants in Bordeaux red grape juice, in the commercial yeast used for its fermentation, and in the produced wine, *J Agric food Chem*, Vol. 48, 400–406. 2000.

30. Guth, H., Identification of character impact odorants of different white wine varieties, *J Agric Food Chem*, Vol. 45, pp. 3022–3026, 1997.

31. Lee, S.-J. and Noble, A.C., Characterization of odor-active compounds in Californian chardonnay wines using GC-olfactometry and GC-mass spectrometry, *J Agric Food Chem*, Vol. 51, 8036–8044, 2003.

32. Baumes, R., Cordonnier, R., Nitz, S., and Drawert, F., Identification and determination of volatile constituents in wines from different vine cultivars, *J Sci Food Agric*, Vol. 37, 927–943, 1986.

33. Wang, Y., Finn, C., and Qian, M.C., Impact of growing environment on Chickasaw blackberry (Rubus L.) aroma evaluated by gas chromatography olfactometry dilution analysis, *J Agric Food Chem*, Vol. 53, 3563–3571, 2005.

34. Peppard, T.L., Volatile flavor constituents of *Monstera deliciosa*, *J Agric Food Chem*, Vol. 40, 257–262, 1992.

35. Hinterholzer, A. and Schieberle, P., Identification of the most odour-active volatiles in fresh, hand-extracted juice of Valencia late oranges by odour dilution techniques, *Flav Fragr J*, Vol. 13, 49–55, 1998.

36. Buettner, A. and Schieberle, P., Characterization of the most odor-active volatiles in fresh, hand-squeezed juice of grapefruit (*Citrus paradisi Macfayden*), *J Agric Food Chem*, Vol. 47, 5189–5193, 1999.

37. Kumazawa, K. and Masuda, H., Identification of potent odorants in different green tea varieties using flavor dilution technique, *J Agric Food Chem*, Vol. 50, 5660–5663, 2002.

38. Czerny, M. and Grosch, W., Potent odorants of raw Arabica coffee: their changes during roasting, *J Agric Food Chem*, Vol. 48, 868–872, 2000.

39. Mick, W. and Schreier, P., Additional volatiles of black tea aroma, *J Agric Food Chem*, Vol. 32, 924–929, 1984.

40. Fan, W. and Qian, M.C., Identification of aroma compounds in Chinese Yanghe Daqu liquor by normal phase chromatography fractionation followed by gas chromatography/olfactometry, *Flav Fragr J*, 21, 333–342, 2006.

41. Ledauphin, J., Saint-Clair, J.-F., Lablanquie, O., Guichard, H., Founier, N., Guichard, E., and Barillier, D., Identification of trace volatile compounds in freshly distilled Calvados and Cognac using preparative separations coupled with gas chromatography-mass spectrometry, *J Agric Food Chem*, Vol. 52, 5124–5134. 2004.

42. Benn, S.M. and Peppard, T.L., Characterization of Tequila flavor by instrumental and sensory analysis, *J Agric Food Chem*, Vol. 44, 557–566, 1996.

43. Phongtonkulphanit, V., Qian, M., and Wang, Y., Normal phase fractionation and GC/O analysis of "Marion" blackberry aroma, presented at IFT Annual Meeting, July 15–20, 2005. New Orleans, LA.

44. Culleré, L., Aznar, M., Cacho, J., and Ferreira, V., Fast fractionation of complex organic extracts by normal-phase chromatography on a solid-phase extraction

polymeric sorbent: optimization of a method to fractionate wine flavor extracts, *J Chromatogr A*, Vol. 1017, 17–26, 2003.

45. López, R., Aznar, M., Cacho, J., and Ferreira, V., Determination of minor and trace volatile compounds in wine by solid-phase extraction and gas chromatography with mass spectrometric detection, *J Chromatogr A*, Vol. 966, 167–177, 2002.

46. Anderson, D.J., High-performance liquid chromatography (advances in packing materials), *Anal Chem*, Vol. 67, 475R–486R, 1995.

47. Scott, R.P.W., *Liquid Chromatography for the Analyst*, Series: Chromatographic Science Series, Ed., Cazes J., Vol. 67, New York: Marcel Dekker, 1994.

48. Parliment, T.H., Concentration and fractionation of aromas on reverse-phase adsorbents, *J Agric Food Chem*, Vol. 29, 836–841, 1981.

49. Unger, K.K. and Weber, E., *A Guide to Practical HPLC*, Darmstadt, Germany: GIT Verlag, 1999.

50. Christie, W.W., Solid-phase extraction columns in the analysis of lipids, in *Advances in Lipid Methodology*, Christie W.W, Ed., Ayr, Scotland: Oily Press, 1992, pp. 1–17.

51. Supelco, Guide to Solid Phase Extraction, *Bulletin 910*, 1998.

52. Harland, C.E., *Ion Exchange: Theory and Practice*, 2nd ed., Cambridge: Royal Society of Chemistry, 1994.

53. The Columbia Encyclopedia (Online), published by Columbia University Press, Copyright 2001–2005, cited July 2005, http://www.bartleby.com/65/ch/chromato.html.

54. McCarthy, R.D. and Duthie, A.H., A rapid quantitative method for the separation of free fatty acids from other lipids, *J Lipid Res*, Vol. 3, 117–119, 1962.

55. Nieuwenhof, F.F.J. and Hup, G., Gas-chromatographic determination of free fatty acids in cheese, *Neth Milk Dairy J*, Vol. 25, 175–182, 1971.

56. Woo, A.H. and Lindsay, R.C., Method for the routine quantitative gas chromatographic analysis of major free fatty acids in butter and cream, *J Dairy Sci*, Vol. 63, 1058–1064, 1980.

57. Woo, A.H. and Lindsay, R.C., Development and characterization of an improved silicic acid-KOH arrestant column for routine quantitative isolation of free fatty acids, *JAOCS*, Vol. 57, 414–416, 1980.

58. de Jong, C. and Badings, H.T., Determination of free fatty acids in milk and cheese, procedures for extraction, clean up, and capillary gas chromatographic analysis, *J High Res Chromatogr*, Vol. 13, 94–98, 1990.

59. Deeth, H., Fitz-Gerald, C., and Snow, A., A gas chromatographic method for the quantitative determination of free fatty acids in milk and milk products, *N Z J Dairy Sci*, Vol. 18, 13–20, 1983.

60. Hornstein, I., Alford, J., Elliot, L., and Crowe, P., Determination of free fatty acids in fat, *Anal Chem*, Vol. 32, 540–542, 1960.

61. Needs, E.C., Ford, G.D., Owen, A.J., Tuckley, B., and Anderson, M., A method for the quantitative determination of individual free fatty acids in milk by ion exchange resin adsorption and gas-liquid chromatography, *J Dairy Res*, Vol. 50, 321–329, 1983.

62. Spangelo, A., Karijord, O., Svenson, A., and Abrahamsen, R.K., Determination of individual free fatty acids in milk by strong anion-exchange resin and gas chromatography, *J Dairy Sci*, Vol. 69, 1787–1792, 1986.

63. McNeill, G.P., O'Donoghue, A., and Connolly, J.F., Quantification and identification of flavour components leading to lipolytic rancidity in stored butter, *Ir J Food Sci Technol*, Vol. 10, 1–10, 1986.

64. McNeill, G.P. and Connolly, J.F., A method for the quantification of individual free fatty acids in cheese: Application to ripening of Cheddar-type cheeses, *Ir J Food Sci Technol*, Vol. 13, 119–128, 1989.

65. Ciucanu, I. and Kerek, F., Selective gas chromatographic determination of the free fatty acids from lipid mixtures, *J Chromatogr*, Vol. 257, 101–106, 1983.

66. Drake, M.A., Gerard, P.D., Kleinhenz, J.P., and Harper, W.J., Application of an electronic nose to correlate with descriptive sensory analysis of aged Cheddar cheese, *Lebensm-Wiss U-Technol*, Vol. 36, 13–20, 2003.

67. Frank, D.C., Owen, C.M., and Patterson, J., Solid phase microextraction (SPME) combined with gas-chromatography and olfactometry-mass spectrometry for characterization of cheese aroma compounds, *Lebensm-Wiss U-Technol*, Vol. 37, 139–154, 2004.

68. Zehentbauer, G. and Reineccius, G.A., Determination of key aroma components of Cheddar cheese using dynamic headspace dilution assay, *Flav Fragr J*, Vol. 17, 300–305, 2002.

69. Qian, M. and Reineccius, G.A., Importance of free fatty acids in Parmesan cheese, in *Heteroatomic Aroma Compounds*, Reineccius, G.A. and Reineccius, T.A., Eds., Washington, D.C.: American Chemical Society, 2002, pp. 243–256.

70. Qian, M. and Reineccius, G.A., Quantification of aroma compounds in Parmigiano Reggiano cheese by a dynamic headspace gas chromatography-mass spectrometry technique and calculation of odor activity value, *J Dairy Sci*, Vol. 86, 770–776, 2003.

71. Chavarri, F., Virto, M., Martin, C., Najera, A., Santisteban, A., Barron, L., and DeRenobales, M., Determination of free fatty acids in cheese: comparison of two analytical methods, *J Dairy Res*, Vol. 64, 445–452, 1997.

72. Bateman, H.G. and Jenkins, T.C., Method for extraction and separation by solid phase extraction of neutral lipid, free fatty acids, and polar lipid from mixed microbial cultures, *J Agric Food Chem*, Vol. 45, 132–134, 1997.

73. Vaghela, M.N. and Kilara, A., A rapid method for extraction of total lipids from whey protein concentrates and separation of lipid classes with solid phase extraction, *JAOCS*, Vol. 72, 1117–1121, 1995.

74. Kaluzny, M.A., Duncan, L.A., Merritt, M.V., and Epps, D.E., Rapid separation of lipid classes in high yield and purity using bonded phase columns, *J Lipid Res*, Vol. 26, 135–140, 1985.

75. Qian, M., Bloomer, S., and Nelson, C., Contribution of free fatty acids to the flavor of various cheeses, presented at 92nd AOCS Annual Meeting and Expo, May 13–16, 2001, Minneapolis, MN.

76. Fang, Y. and Qian, M.C., Pinot noir wine aroma analysis by sequential ion exchange and normal phase chromatography-gas chromatography/olfactometry, presented at ASEV Annual Meeting, June 30–July 2, 2004, San Diego, CA.

77. Tominaga, T., Murat, M.-L., and Dubourdieu, D., Development of a method for analyzing the volatile thiols involved in the characteristic aroma of wines made from *Vitis vinifera* L. Cv. Sauvignon Blanc, *J Agric Food Chem*, Vol. 46, 1044–1048, 1998.

78. Jocelyn, P., Spectrophotometric assay of thiols, in *Sulfur and Sulfur Amino Acids*, Jakoby, W. and Griffith, O., Eds., Orlando, FL: Academic Press, 1987, pp. 44–67.

79. Tominaga, T., Darriet, P., and Dubourdieu, D., Identification de l'acetate de 3-mercaptohexanol, composé á forte odeur de buis, intervenant dans l'arôme des vins de Sauvignon, *Vitis*, Vol. 35, 207–210, 1996.

80. Kleinhenz, J.P., Medium and higher molecular weight volatile thiols in aged cheddar cheese and their relation to flavor, Ph.D. dissertation, Ohio State University: Columbus, OH, 2003.

81. Davis, J.M. and Giddings, J.C., Statistical method for estimation of number of components from single complex chromatograms: application to experimental chromatograms, *Anal Chem*, Vol. 57, 2178–2182, 1985.

82. Pryde, A. and Gilbert, M.T., *Applications of High Performance Liquid Chromatography*, London: Chapman and Hall, 1979.

83. Ferreira, V., Hernandez-Orte, P., Escudero, A., Lopez, R., and Cacho, J., Semi-preparative reversed-phase liquid chromatographic fractionation of aroma extracts from wine and other alcoholic beverages, *J Chromatogr A*, Vol. 864, pp. 77–88. 1999.

84. Escudero, A., Ferreira, V., and Cacho, J., Isolation and identification of odorants generated in wine during its oxidation: a gas chromatography-olfactometric study, *Eur Food Res Technol*, 2000, Vol. 211, pp. 105–110.

85. Aznar, M., López, R., Cacho, J.F., and Ferreira, V., Identification and quantification of impact odorants of aged red wines from Rioja. GC-Olfactometry, quantitative GC-MS, and odor evaluation of HPLC fractions, *J Agric Food Chem*, 2001, Vol. 49, 2924–2929.

86. Barron, D., HPLC using diol-bonded silica, an alternative to silica gel in the prefractionation of aroma extracts, *Z Lebensm Unters Forsch*, Vol. 193, 454–459, 1991.

87. Schmidt-Traub, H., *Preparative Chromatography of Fine Chemicals and Pharmaceutical Agents*, Weinheim, Germany: Wiley-VCH Verlag, 2005.

88. Scott, R.P.W., *Gas Chromatography*, Chrom-Ed Series: The Instant Chromatography Reference Library, published by Library4Science, Copyright 2002–2003, cited August 2005, http://www.chromatography-online.org/3/contents.html.

89. Engel, W. and Schieberle, P., Structural determination and odor characterization of N-(2-mercaptoethyl)-1,3-thiazolidine, a new intense popcorn-like-smelling odorant, *J Agric Food Chem*, Vol. 50, 5391–5393, 2002.

90. Rogerson, F.S.S., Castro, H., Fortunato, N., Azevedo, Z., Macedo, A., and de Freitas, V.A.P., Chemicals with sweet aroma descriptors found in Portuguese wines from the Douro region: 2,6,6-trimethylcyclohex-2-ene-1,4-dione and diacetyl, *J Agric Food Chem*, Vol. 49, 263–269, 2001.

91. de Freitas, V.A.P., Ramalho, P.S., Azevedo, Z., and Macedo, A., Identification of some volatile descriptors of the rock-rose-like aroma of fortified red wines from Douro demarcated region, *J Agric Food Chem*, Vol. 47, 4327–4331, 1999.

92. Specht, K. and Baltes, W., Identification of volatile flavor compounds with high aroma values from shallow-fried beef, *J Agric Food Chem*, Vol. 42, 2246–2253, 1994.

93. Parliment, T.H., Clinton, W., and Scarpellino, R., trans-2-Nonenal: coffee compound with novel organoleptic properties, *J Agric Food Chem*, Vol. 21, 485–487, 1973.

94. Holscher, W., Vitzthum, O.G., and Steinhart, H., Prenyl alcohol — source for odorants in roasted coffee, *J Agric Food Chem*, Vol. 40, 655–658. 1992.

95. Shellie, R. and Marriott, P., Opportunities for ultra-high resolution analysis of essential oils using comprehensive two-dimensional gas chromatography: a review, *Flav Fragr J*, Vol. 18, 179–191, 2003.

96. Merritt, C., Application in flavor research, in *Preparative Gas Chromatography*, Zlatkis, A. and Pretorius, V., Eds., New York: Wiley-Interscience, 235–276, 1971.

97. Phillips, J.B. and Xu, J., Comprehensive multi-dimensional gas chromatography (review), *J Chromatogr A*, Vol. 703, 327–334, 1995.

98. Bushey, M.M. and Jorgenson, J.W., Automated instrumentation for comprehensive two-dimensional high-performance liquid chromatography of proteins, *Anal Chem*, Vol. 62, 161–167, 1990.

99. Pursch, M., Sun, K., Winniford, B., Cortes, H., Weber, A., McCabe, T., and Luong, J., Modulation techniques and applications in comprehensive two-dimensional gas chromatography (GC × GC), *Anal Bioanal Chem*, Vol. 373, 356–367, 2002.

100. Phillips, J.B and Beens, J., Comprehensive two-dimensional gas chromatography: a hyphenated method with strong coupling between the two dimensions (review), *J Chromatogr*, Vol. 856, 331–347, 1999.

101. Bertsch, W., Two-dimensional gas chromatography. Concepts, instrumentation, and application — Part 1: Fundamentals, conventional two-dimensional gas chromatography, selected applications, *J High Res Chromatogr*, Vol. 22, 647–665, 2000.

102. Schomburg, G., Husmann, H., Hubinger, E., and Konig, W., Multidimensional capillary gas chromatography-enantiomeric separations of selected cuts using a chiral second column, *J High Res Chromatagr*, Vol. 7, 404–410, 1984.

103. Rijks, J.P.E.M. and Rijks, J.A., Programmed cold sample introduction and multidimensional preparative capillary gas chromatography, *J High Res Chromatagr*, Vol. 13, 261–266, 1990.

104. Darriet, P., Pons, M., Henry, R., Dumont, O., Findeling, V., Cartolaro, P., Calonnec, A., and Dubourieu, D., Impact odorants contributing to the fungus type aroma from grape berries contaminated by powdery mildew (*Uncinula necator*): incidence of enzymatic activities of the yeast *Saccharomyces cerevisiae*, *J Agric Food Chem*, Vol. 50, 3277–3282, 2002.

105. Kempfert, K.D., Evaluation of apparent sensitivity enhancement in GC/FTIR using multidimensional GC techniques, *J Chromatogr Sci*, Vol. 27, 63–70, 1989.

106. Gordon, B.M., Uhrig, M.S., Borgerding, M.F., Chung, H.L., Coleman, W.M., Elder, J.F., Jr., Giles, J.A., Moore, D.S., Rix, C.E., and White, E.L., Analysis of flue-cured tobacco essential oil by hyphenated analytical techniques, *J Chromatogr Sci*, Vol. 26, 174–180, 1988.

107. Scott, R.P.W., *Principles and Practice of Chromatography*, Chrom-Ed Series: The Instant Chromatography Reference Library, published by Library4Science, Copyright 2002–2003, cited August 2005, http://www.chromatography-online.org/1/contents.html.

108. Full, G., Winterhalter, P., Schmidt, G., Herion, P., and Schreier, P., MDGC-MS: a powerful tool for enantioselective flavor analysis, *J High Res Chromatagr*, Vol. 16, 642–644, 1993.

109. Shellie, R., Mondello, L., Marriott, P., and Dugo, G., Characterization of lavender essential oils by using gas chromatography-mass spectrometry with correlation of linear retention indices and comparison with comprehensive two-dimensional gas chromatography, *J Chromatogr A*, Vol. 970, 225–234, 2002.

110. Adahchour, M., Brandt, M., Baier, H.-U., Vreuls, R.J.J., Batenburg, A.M., and Brinkman, U.A.T., Comprehensive two-dimensional gas chromatography coupled to a rapid-scanning quadrupole mass spectrometer: principles and applications, *J Chromatogr A*, Vol. 1067, 245–254, 2005.

111. van Mispelaar, V.G., Tas, A.C., Smilde, A.K., Schoenmakers, P.J., and van Asten, A.C., Quantitative analysis of target components by comprehensive two-dimensional gas chromatography, *J Chromatogr A*, Vol. 1019, 15–29, 2003.
112. Zhu, S., Lu, X., Dong, L., Xing, J., Su, X., Kong, H., Xu, G., and Wu, C., Quantitative determination of compounds in tobacco essential oils by comprehensive two-dimensional gas chromatography coupled to time-of-flight mass spectrometry, *J Chromatogr A*, Vol. 1086, 107–114, 2005.
113. Dimandja, J.-M.D., Stanfill, S.B., Grainger, J., and Patterson, D.G., Application of comprehensive two-dimensional gas chromatography (GC × GC) to the qualitative analysis of essential oils, *J High Res Chromatogr*, Vol. 23, 208–214, 2000.
114. Adahchour, M., Beens, J., Vreuls, R.J.J., Batenburg, A.M., Rosing, E.A.E., and Brinkman, U.A.T., Application of solid-phase microextraction and comprehensive two-dimensional gas chromatography (GC × GC) for flavor analysis, *Chromatographia*, Vol. 55, 361–367, 2002.
115. Shellie, R., Marriott, P., and Cornwell, C., Application of comprehensive two-dimensional gas chromatography (GC × GC) to the enantioselective analysis of essential oils, *J Sep Sci*, Vol. 24, 823–830, 2001.
116. Ryan, D., Shellie, R., Tranchida, P., Casilli, A., Mondello, L., and Marriott, P., Analysis of roasted coffee bean volatiles by using comprehensive two-dimensional gas chromatography-time-of-flight mass spectrometry, *J Chromatogr A*, Vol. 1054, 57–65, 2004.
117. Mondello, L., Casilli, A., Tranchida, P.Q., Dugo, P., Costa, R., Festa, S., and Dugo, G., Comprehensive multidimensional GC for the characterization of roasted coffee beans, *J Sep Sci*, Vol. 27, 442–450, 2004.
118. Adahchour, M., Wiewel, J., Verdel, R., Vreuls, R.J.J., and Brinkman, U.A.T., Improved determination of flavour compounds in butter by solid-phase (micro)extraction and comprehensive two-dimensional gas chromatography, *J Chromatogr A*, Vol. 1086, 99–106, 2005.
119. Adahchour, M., van Stee, L.L.P., Beens, J., Vreuls, R.J.J., Batenburg, M.A., and Brinkman, U.A.T., Comprehensive two-dimensional gas chromatography with time-of-flight mass spectrometric detection for the trace analysis of flavour compounds in food, *J Chromatogr A*, Vol. 1019, 157–172, 2003.
120. Williams, A., Ryan, D., Guasca, A.O., Marriott, P., and Pang, E., Analysis of strawberry volatiles using comprehensive two-dimensional gas chromatography with headspace solid-phase microextraction, *J Chromatogr B*, Vol. 817, 97–107, 2005.

chapter six

Solid phase dynamic extraction: a technique for extracting more analytes from samples

Ingo Christ, Ulrike B. Kuehn, and Ken Strassburger

Contents

I. Introduction to solid phase dynamic extraction

A relatively new compromise between solid-phase microextraction (SPME) and stir bar sorptive extraction (SBSE) is generating significant interest in the U.S. food and flavor industry. Known as solid phase dynamic extraction (SPDE), the technique and associated tools were developed in 2000 by CHROMTECH GmbH of Idstein, Germany. Essentially, SPDE is an inside-needle technique for vapor and liquid sampling, which is as easy to apply and as reproducible as static headspace (S-HS) but with increased capacity and reduced sample handling. SPDE offers several inherent advantages over SPME and SBSE. Although SPME provides high extraction speeds and stability, one of its chief limitations is its reduced concentration capability, which is mainly due to the small volume of polymer that coats the fiber. In an SPME 100-μm fiber, there is only about 0.6 μl of PDMS, whereas SPDE contains about 4.5 μl of the polymer. Other disadvantages of SPME include the fragility of its fused-silica, its unprotected stationary phase coating, and the limited flexibility of its surface area. Several attempts have been made to overcome these disadvantages. Most notably, in 1997, Ralf Eisert and Janusz Pawliszyn successfully introduced in-tube SPME–LC, in which sampling is conducted through an open tubular fused-silica capillary column.

SBSE was first introduced by Pat Sandra of the University of Ghent, Belgium. SBSE is characterized by a significant concentration capacity, enabled by a high volume of PDMS coating (from 25 to 200 μl) on magnetic stir bars. Thus far, however, only PDMS coating has been available for SBSE, and the technique requires long extraction times due to the small surface area.

SPDE combines all of the traditional advantages of SPME with SBSE's increased capacity, and it offers a large phase selection as well. In fact, SPDE approaches the sensitivity of purge and trap technology but with greater automation and simplified interfacing to the GC inlet. Also the inlet is not blocked by additional hardware, so that liquid and headspace can still be performed using the same inlet.

A. Principles of SPDE

$$J \equiv \frac{1}{A}\frac{dn_i}{dt} = -D_i\frac{dc_i}{dx} = -D_L\frac{dc_L}{dx} \tag{6.1}$$

In this equation, "J" represents the mass flux of the analyte from the SPDE phase to the sample matrix; A represents the surface area of the SPDE phase; dn_i represents the amount of analyte desorbed from the SPDE phase during time period dt; D_i and D_L represent diffusion coefficients of the analyte in the SPDE coating and the sample matrix; dc_i represents the concentration in the

SPDE polymer; and dc_L represents the concentration in the boundary layer. A linear concentration gradient in the polymer coating and boundary layer is assumed:

$$\frac{d\,n_i}{A\cdot dt} = \frac{D_i}{\delta_i}(c_i - \acute{c}_i) = -\frac{D_L}{\delta_L}(c_L - \acute{c}_L) \tag{6.2}$$

Here, δ_i and δ_L represent the thickness of the SPDE coating and the boundary layer; c_i represents the concentration of the analyte in the coating on the steel needle, compared to the concentration on the surface $\acute{c}_i$; $\acute{c}_i$ represents the concentration of the analytes in the coating on its surface; c_L represents the concentration in the boundary layer; and $\acute{c}_L$ represents the concentration in the sample matrix (see Figure 6.1).

The mass transfer coefficients of the analyte in the polymer coating k_i and the analyte in the boundary layer k_L are defined as $k_i = {}^{D_i}\!/\!_{\delta_i}$ and $k_L = {}^{D_L}\!/\!_{\delta_L}$:

$$\frac{d\,n_i}{A\cdot dt} = -k_i(c_i - \acute{c}_i) = -k_L(c_L - \acute{c}_L) \tag{6.3}$$

It is assumed that an immediate partition equilibrium exists at the interface of the polymer coating and the boundary layer:

$$K = \frac{c_i}{c_L} - \acute{c}_L = \frac{c_i}{K} \tag{6.4}$$

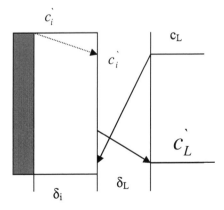

Figure 6.1 Schematic of the extraction process and the various concentrations close to the SPDE coating.

Here, K represents the distribution constant of the analyte between the polymer coating and the sample matrix. In the bulk of aqueous solution:

$$c_L = \frac{q}{V_S} \tag{6.5}$$

V_s describes the sample volume. Applying Equation 6.4 and Equation 6.5 to Equation 6.3 results in:

$$-k_i(c_i - \dot{c}_i) = -k_L\left(\frac{q}{V_S} - \frac{c_i}{K}\right) - c_i - \dot{c}_i = \frac{k_L\left(\dfrac{q}{V_S} - \dfrac{c_i}{K}\right)}{k_i} \tag{6.6}$$

As there is a linear concentration gradient in the polymer layer:

$$q_0 - q = V_i\frac{c_i + \dot{c}_i}{2} - c_i + \dot{c}_i = \frac{2(q_0 - q)}{V_i} \tag{6.7}$$

Here, q_0 is the amount of analyte initially loaded onto the polymer coating before its exposure to the sample matrix.

Combining Equation 6.6 and Equation 6.7 results in:

$$2c_i = \frac{k_L\left(\dfrac{q}{V_S} - \dfrac{c_i}{K}\right)}{k_i} + \frac{2(q_0 - q)}{V_i} - c_i = \frac{K[k_L V_i q = 2(q_0 - q)k_i V_S]}{V_S V_i(2k_i K + k_S)} \tag{6.8}$$

Applying Equation 6.8 to the right side of Equation 6.6 results in:

$$c_i - \dot{c}_i = \frac{k_S}{k_i}\left[\frac{q}{V_S} - \frac{k_S V_i q + 2(q_0 - q)k_i V_S}{V_S V_i(2k_i K + k_S)}\right] = \frac{2q(KV_i k_S + V_S k_S)}{2k_i KV_S V_i + k_S V_S V_i} - \frac{2q_0 k_S}{2k_i KV_i + k_S V_i} \tag{6.9}$$

Applying Equation 6.9 to the right side of Equation 6.7 results in:

$$\frac{dq}{A \cdot dt} = -k_i\left[\frac{2q(KV_i k_S + V_S k_S)}{2k_i KV_S V_i + k_S V_S V_i} - \frac{2q_0 k_S}{2k_i KV_i + k_S V_i}\right] \tag{6.10}$$

Let

$$a = \frac{2Ak_i(KV_ik_S) + V_Sk_S}{2k_iKV_SV_i + k_SV_SV_i} \tag{6.11}$$

$$b = \frac{2q_0Ak_Sk_i}{2k_iKV_i + k_SV_i} \tag{6.12}$$

Then Equation 6.10 simplifies to:

$$q^1 + aq = b \tag{6.13}$$

The general solution to Equation 6.13 is:

$$q \exp \int a\, dt = \int b\left(\exp \int a\, dt\right) dt + Z \tag{6.14}$$

$$q \exp(at) = \frac{b}{a}(\exp(at) - 1) + Z \tag{6.15}$$

The boundary condition to Equation 6.15 is $t = 0$, $q = 0$. Then $Z = 0$. Equation 6.15 results in:

$$q = \frac{b}{a} \cdot [1 - \exp(-at)] \tag{6.16}$$

Equation 6.12 divided by Equation 6.11 results in:

$$\frac{b}{a} = \frac{V_S}{KV_i + V_S} \cdot q_0 \tag{6.17}$$

Applying Equation 6.17 to Equation 6.16 results in:

$$q = q_0[1 - \exp(-at)] \cdot \frac{V_S}{KV_i + V_S} \tag{6.18}$$

If $V_S \geq KV_i$, which would be the case with any real-life sample, Equation 6.18 can be simplified to:

$$q = q_0[1 - \exp(-at)] \tag{6.19}$$

Let $Q = q_0 - q$, where Q is the amount of the analyte remaining on the polymer coating after its exposure to the sample matrix for sampling time t, then:

$$Q = q_0 \exp(-at) \tag{6.20}$$

Parameter *a* describes the kinetics of desorption of an analyte into the SPDE polymer.

B. Instrumentation

When discussing SPDE, it is always assumed that the Combi PAL autosampler is being used. The autosampler is manufactured by CTC Analytics, Zwingen, Switzerland. It allows the necessary flexibility to choose all important parameters, and it provides high reproducibility including needle cleanup and conditioning. In addition, most other known sample introduction techniques can be performed with the autosampler, including liquid, headspace SPME, SBSE, thermal desorption, and split or splitless injections.

SPDE equipment (syringes with attached SPDE needles and an SPDE gas station) can be installed on a CTC Combi PAL autosampler, which, in turn, is assembled on a GC or GC-MS system (see Figure 6.2). The autosampler includes a Single Magnet Mixer oven with one heated vial position (manufactured and distributed by CHROMTECH GmbH/CHROMSYS LLC) or a 6-position agitator. All SPDE sampling steps are automatically controlled by the Combi PAL software. The SPDE needle (50 × 0.8 mm, I.D. 0.53 mm, conical needle tip with side port) is coated on the inside with a polymer such as PDMS containing 10% activated carbon and another dozen other phases. This needle is assembled onto a 2.5-ml gas-tight syringe with a side port for gas flushing (manufactured by Hamilton, Darmstadt, Germany). The gas station and syringe are connected to helium

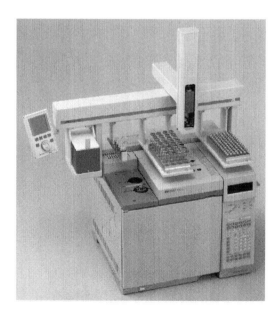

Figure 6.2 GC-system with CTC Combi PAL autosampler and SPDE equipment.

for flushing and are controlled by the autosampler. The gas station supplies a fixed volume of helium for desorption. The side port of the syringe cannot be used for desorption because it has no pressure regulator.

C.   Technical aspects of SPDE

A gas-tight 2.5-ml syringe is equipped with a special needle that is coated on the inside with an extraction phase. The phase can be applied in variable thicknesses.

The adsorption process is best described as follows: A liquid or headspace sample is drawn into the 2.5-ml syringe, thereby adsorbing analytes onto the stationary phase. A distribution balance is reached between the liquid sample matrix and the active phase (see Equation 6.19). The analytes are concentrated onto the phase by repeatedly moving the plunger up and down, thus forcing the headspace or fluid through the needle. Fast plunger speeds ensure a quick exchange of sample near the active phase. After adsorption, the syringe picks up carrier gas from the fiber bakeout station (if necessary, the station also is used to dry the coating and syringe). Then, the autosampler moves the syringe over to the hot GC inlet, and the analytes are desorbed. Additional gas flow from the syringe forces the analytes into the inlet, thereby ensuring complete desorption and sharp peak shapes. This technique can be used with splitless flow for maximum sensitivity. For volatile compounds a cryogenic oven or on-column cooling helps increasing the sensitivity due to improved refocusing on the column.

Depending on the sample and active coating, SPDE syringes can be used for 500 to 1000 injections. The highest amount thus far is 3000 injections using a single syringe with no degradation. Individual sample matrices and desorption temperatures essentially determine syringe longevity. Low-level headspace extractions represent the ideal sample and can therefore offer the longest lifetime, whereas aqueous liquids with particles likely will prevent the SPDE syringe from having a very long lifetime. In any event, the sturdy steel syringe needle with thicker films prevents the possible problem of the mechanical failure of SPME fibers and needles (see Figure 6.3).

D.   Influence of parameters

1.   Extraction parameters

After the equilibration time, the headspace in a sealed vial reaches its equilibrium with dissolved volatile or semi-volatile compounds of the sample. With each aspiration stroke during the extraction, the balance constantly changes as the polymer dissolves the volatile analytes, while the reduction in concentration of the headspace allows the analytes to reevaporate from the sample (solution or solid) from the bottom of the vial.

Several parameters are variable while others are preset. For the extraction process, the focus is on incubation temperature, extraction strokes (including aspiration speed and dispension speed).

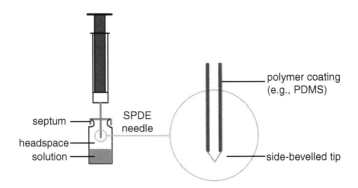

Figure 6.3 SPDE needle.

   a.   Temperature.   The influence of different temperatures on HS-SPDE recovery of a sample needs to be evaluated under standardized conditions. One important consideration for an ideal temperature is that a compromise needs to be found for the varying polarity and volatility of the components of both sample and real-world matrices. Secondly, some of the polymers work best at certain temperature ranges because the dissolution process can be reversed at higher temperatures. This is especially true for compounds with a low boiling point. Mixing plays a role only when using a liquid or liquefied sample with a reasonable viscosity. In this case, it might prove important to have a decent mixing process because the aqueous layer might otherwise prevent any exchange from the sample to the headspace. Also, the choice of stir bar is important in case highly viscous or somewhat dirty samples are used.

   b.   Extraction strokes.   Each fill stroke uses the fixed volume of 1000 µl. This value has been set by the manufacturer to reduce the risk of a leakage through a vacuum inside the vial. Settable parameters include the speed of each aspiration and dispension. The speed determines the thickness of the boundary layer above the coating where there is laminar flow. It also determines the time that a specific compound stays within the layer. A good compromise needs to be found in order to enable a fast exchange of analytes within the boundary layer, so that there is always a fresh supply of new analytes and that they stay long enough to be attracted to the polymer. Another factor is that the aspiration might decrease the concentration of the analytes so significantly that during dispension, not many analytes will be left to dissolve, so the process will actually be reversed. These phenomena are adequately described in the previous mathematical treatment. However, when dealing with a complex food, flavor, or fragrance matrix, certain compromises need to be made.
   Another important consideration is the durability of the plunger. When very fast speeds are used, the plunger will wear out quickly. Typically, a plunger's lifetime does not exceed 50,000 cycles. Its lifetime is also temperature dependent. Typically, at a temperature of 60°C or less, no problems have been attributed to plunger wear. A replacement plunger can easily be acquired.

2. Desorption parameters

Regardless of the extraction parameters, it is always useful to have a thick film column or some type of cryogenic cooling installed. These parameters maintain the concentration of analyte per volume of gas for sharp peak shapes and maximum detector sensitivity. A fast column or a high oven temperature can also aid refocusing on the head of the column. A trap or retention gap with a thick film might also prove useful. A typical configuration may include subambient cooling of the entire GC oven or an on-column cryo-focusing unit such as provided by Brechbühler, Inc., SIS or SGE.

Before discussing the parameters of the SPDE desorption, it is important to understand the technique. Because it uses heat for desorption, it also uses carrier gas to drive out the analytes. In other words, it is a mixed headspace and extraction technique.

The desorption process of SPDE uses the following parameters:

- Predesorption time
- Extraction gas volume
- Plunger speed

The goal of the desorption process is to efficiently desorb all previously dissolved analytes from the polymer and, at the same time, to achieve sharp peak shapes. The latter two parameters are essential to achieving higher sensitivity because both of them have a significant effect on it.

If the desorption speed is too slow, peak broadening might cause fat peaks, whereas too high of a desorption speed might overpressurize the inlet. This overpressurization would result in a similar chromatogram with fat peaks.

The desorption volume corresponds directly to desorption speed in such a way that it is limited by the inlet liner volume. When more volume is available, it allows more gas without significant overpressurization.

In other words, the liner has a very significant effect on the process. A standard split or splitless liner with a 4 mm I.D. and a recess on the bottom is the preferred type to use, because it allows the most volume and provides great sensitivity. In recent times, the so-called Uniliner was introduced, and it has proven to be helpful when splitless transfers are used. The Uniliner seals tightly around the column and works similarly to an on-column injector. It permits maximum sensitivity because it transfers everything out of the liner into the column and still allows great peak shapes — especially when used with one of the cryo-focusing techniques already mentioned. An additional benefit is that the same injector configuration can be used for normal liquid injections; that is, the system is not necessarily headspace only and will allow the injection of retention time standards or recovery studies.

3. Summary of parameters

Depending on volatile analyte and matrix, changing a parameter may have a very different outcome. To better understand the changes in concentration, we illustrate an example from Dr. Carlo Bicchi of the University of Turin, Italy [1].

He studied the effects of SPDE on different parameters. He used a standard of β-pinene (BP 165 to 168°C), isoamyacetate (BP 142°C) and linaool (BP 199°C). The three compounds provided a fairly large boiling point range. The effects are shown in Table 6.1.

Using the common PDMS-coating as a stationary phase, β-pinene and isoamyl acetate are best extracted by utilizing quick parameters with a large desorption volume, because linalool requires a higher extraction temperature and longer extraction time. Lately, Strassburger has demonstrated that a Carboxen™ (activated carbon) phase is much less prone to these discriminations.

E. Choice of polymer coating for the stationary phase

Using SPDE, analytes are concentrated onto a 50 μm film of extraction coating. The most common phase is polydimethylsiloxane (PDMS) and Carboxen (10%), which are coated on the inside wall of a stainless steel needle (56 mm length) in a 2.5-ml gas tight syringe. A long version of 72 mm is also available. Many other phases — including ones with higher polarity — are available as well. Particularly interesting is the option of applying new polymers such as developmental ones. In those cases, PDMS is often used as an undercoating because it bonds very well with the stainless steel needle. Currently available coatings and related applications can be found in Table 6.2.

When SPDE is used for headspace sampling (HS-SPDE), a fixed volume of the sample's headspace is aspirated with the gas-tight syringe an appropriate number of times, and an analyte amount suitable for reliable GC or GC-MS analysis accumulates in the polymer coating of the needle wall. Studies show that HS-SPDE is a successful technique for HS-sampling with high concentration capability, good repeatability, and reproducibility, including in comparisons with HS-SPME.

Table 6.1 Percent Changes in Peak Areas of β-Pinene, Isoamylacetate, and Linalool as a Function of Changes in SPDE Parameters

SPDE parameter	Change	β-Pinene, isoamylacetate	Linalool
Extraction temperature	30–70°C	−50% (from peak area 80/60)	+ 600% (to 30 from 5)
Extraction cycles	50–150	−50% (from 35/20)	+ 30% (from 15)
Extraction plunger speed	50–100	−30%/−25% (from maximum with 35/22)	+ 10% (from 15)
Extraction volume	0.5–2 ml	−50%/−30% (from 65/30)	+ 120% (from 10)
Desorption volume	0.5–2 ml	+ 40% + 50% (from 30/12)	+ 150% (from 8)
Desorption plunger speed	10–25	± 5%	+ 15%

Table 6.2 Polymer Coatings for SPDE and Applications

Commercial name	Description	Application	Pol.
			Least Polar ----------------Most Polar
BP1, DB1, HP1	100% Dimethyl polysiloxane (PDMS)	Volatiles	
BP5, DB5, HP5	5% Phenyl methylpolysiloxane, 95% PDMS	Semivolatiles	
PDMS/Carboxen	90% PMDS and activated carbon or carboxen	Volatiles and polar compounds	
BP10, DB1701, HP1701	14% Cyanopropylphenyl polysiloxane	Pesticides	
BP225, DB225, HP225	50% Cyanopropylphenyl polysiloxane	Fatty esters, aldehydes	
BP20 Wax, DB Wax, HP Innowax	Polyethylene glycol (wax)	Alcohols, esters, and aldehydes	

F. SPDE extraction cooler for highly volatile compounds

SPDE has proven particularly interesting for food, flavor, fragrance, and odor chemistry, because it offers significantly more sensitivity for compounds with low volatility.

In Europe, a new device has been patented for SPDE cooling: the SPDE Extraction Cooler. It uses a triple peltier cooler to cool the needle 40°C below room temperature. A picture of the SPDE Extraction Cooler is shown in Figure 6.4.

The difference between a standard extraction and one with a cooler is that for the latter, the needle is cooled while the sample and syringe temperature are elevated. This provides maximum extraction performance because the active compounds such as PDMS dissolve better at lower temperatures. This is especially true for very low boiling compounds, which previously were not detectable. Using the example of methyl-1, 1-dimethylethyl ether (MTBE) (BP 55°C), a considerable improvement in recovery occurred with the SPDE Extraction Cooler. A 20 ng/l MTBE (Signal to Noise 45:1 on SIM, Ion 63) was easily detectable in a VOC mix of 18 compounds.

The extraction cooler could conceivably be used for SPME fiber cooling. Strassburger has recently demonstrated that static headspace sampling can be performed using the coated needle setup in the analysis of acetaldehyde. Even though an equilibrium is established according to Equation 6.19 and Equation 6.20, there is negligible absorption after a single syringe stroke and very rapid release in the heated injector.

Figure 6.4 The SPDE extraction cooler is an enhancement that mounts on the single magnet mixer. The SPDE needle protrudes only a few millimeters into the vial.

II.   Applications

A.   Residues of packaging material in beverages

An examination of the components of a beverage normally results in a variety of substances. Not all of them are natural ingredients. Some compounds are byproducts; others derive from packaging. In this study we checked for butyl acrylate and tri-propyleneglycol diacrylate that is commonly found in the ink of outside packaging.

The analysis of a liner for skim milk [2] by SPDE and GC-MS shows an active chromatogram (Figure 6.5). We chose to determine the GC retention times of the above mentioned analytes by preparing standards and analyzing via single ion monitoring (SIM) using *m/z* 55 and *m/z* 73. Now, with a considerably clearer spectrum, it is also possible to adjust the parameters for better performance. This also eliminates the siloxane background, which is found commonly in PDMS coatings and fibers. Figure 6.6 details the SIM analysis.

We were able to split a typical carton package into three different layers: inner polymer layer, paper layer, and outer polymer layer with print. Similar amounts of inner and outer layer were cut into small pieces and stirred with a saturated aqueous solution of sodium chloride at a temperature of 50°C, 50 strokes @ 100 µl/sec. Headspace analysis showed only some diffusion from outer to inner layer in the case of the milk carton (Figure 6.6), but no diffusion into the milk product itself (Figure 6.7 to Figure 6.9). Much of the

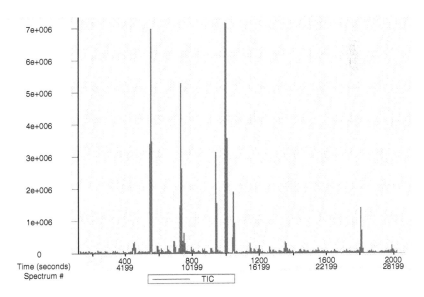

Figure 6.5 The analysis of a milk sample with SPDE and GC-MS shows a variety of compounds.

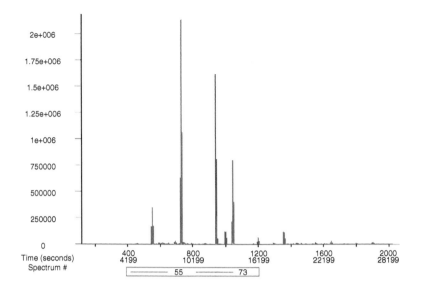

Figure 6.6 The SIM analysis of a milk liner by SPDE and GC-MS showing only *m/z* 55 and *m/z* 73.

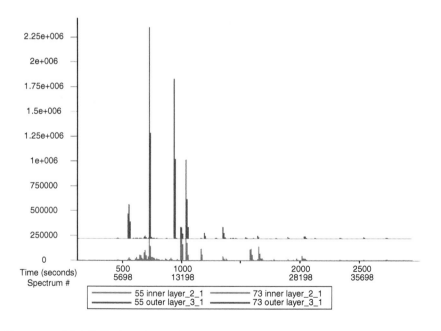

Figure 6.7 GC-MS SIM analysis of inner liner and outer printed liner.

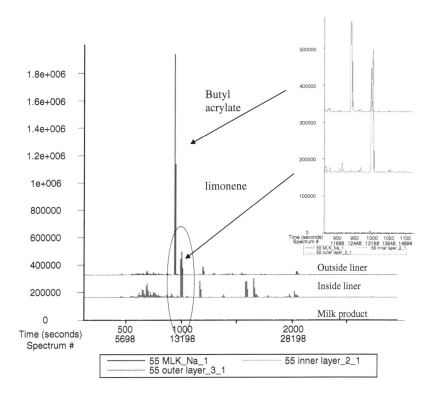

Figure 6.8 GC-MS SIM analysis of milk and liners, m/z = 55, showing a large amount of butyl acrylate in the outer liner layer but is not migrating into the product.

package design today is usually quite good and must consider the product type and solubility of components to deliver the desired profile. For example, limonene is an ubiquitous compound and present in many natural products but would not be desired in milk. It does, however, occur, based on the diet of dairy cows, in cheese and milk in some instances. Figure 6.10 demonstrates high levels of limonene behind the inner carton layer and some, much less amount, in the outer layer, with almost none in the milk product. We cannot determine if limonene was used in the manufacture of the lining, or if it was originally present in the milk product. (A blank run showed no limonene.) We can demonstrate that these poly type liners are very permeable to terpene components, especially the migration from aqueous products, such as milk and juice [3] into the various package liners as shown in Figure 6.11. Here we can see significant flavor compounds in the juice product and in the dissected liner material. This will cause significant quality changes in the shelf products. Figure 6.12 demonstrates key flavor ingredients have migrated into the liner, and Figure 6.13 indicates at least some migration of silanes from package to product. Although it is possible to construct higher density liners, it is not always economically feasible.

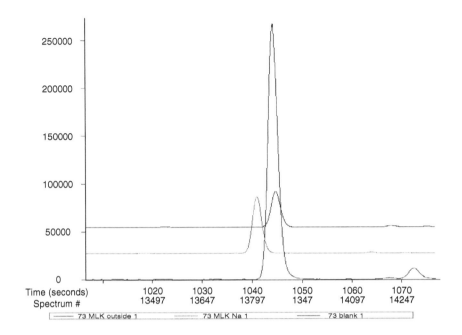

Figure 6.9 Milk/Carton by SPDE GC-MS SIM m/e = 73 showing significant silanes in the outer layer (bottom trace) but very little in the milk (center) or blank run (top). The blank excludes significant silanes as artifacts of the PDMS film.

B. Example: Cola

A comparison of SPDE (solid-phase dynamic extraction) and SPME (solid-phase microextraction) techniques using the example of Coke shows some interesting variations. Generally, SPDE shows higher sensitivity espe-cially for highly volatile compounds. (See Figure 6.14 and Figure 6.15; note that for similar results, SPDE data was measured with a split ratio of 1:100, whereas the split ratio for SPME was 1:10).

The parameters can be adjusted in order to focus on different analytes (see Table 6.3). Changing the number of ejection strokes in the SPDE analysis results in higher yields for more ejection strokes. Also, for the volatiles myrcene, γ-terpinene, borneol, and myristicin, the extraction fill speed and eject speed of 100 μl/sec lead to better sensitivity than a reduced speed of 50 μl/sec does. In the case of cinnamic acid, the slower extraction fill speed leads to slightly higher numbers than those in which longer extraction times distort the profile.

An SPME analysis of cola with lime flavor under similar conditions provides higher relative yields of γ-terpinene, but otherwise the relative yields are similar (Table 6.4). Because of the lower sensitivity of SPME, the GC run was carried out with a lower split ratio of 10:1 instead of 100:1 (Figure 6.15).

In a direct comparison, it can clearly be seen that the higher volatile compounds are better trapped in the SPDE needle due to the higher volume

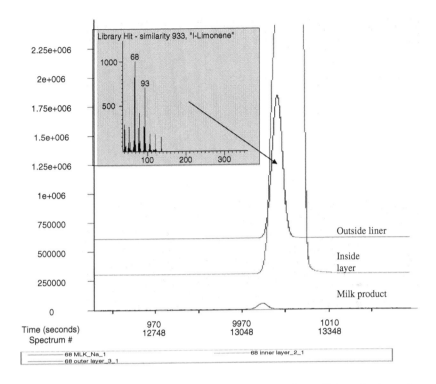

Figure 6.10 The limonene is apparently migrating from the milk into the inner layer but cannot penetrate the outer layer. The display is a SIM presentation of m/e = 68.

of polymer coating, and this is the reason for the higher split ratio of SPDE than that of SPME. Also, besides the overall higher sensitivity, the good recovery of higher volatile compounds with SPDE has to be acknowledged (see Figure 6.16 and Figure 6.17).

C.   Citral with SPDE

One of the many reasons for analyzing food samples is for quality control and the identification of the decomposition products of the fresh and more reactive components. Here, we analyze the performance of an acid stabilization compound (patent pending) to enhance the stability of citral sample in acidic medium. The chromatography, rapid sample preparation, and sensitivity of SPDE are highly suitable for stability testing. Using SPDE, we are able to account for all known off notes arising from citral and terpene oxidation under acidic conditions without resorting to tedious liquid–liquid extraction, with sub-ppm sensitivity (see Figure 6.18 and Figure 6.19).

D.   Volatile fractions of food matrices

Besides studying the optimization of sampling parameters conditioning HS-SPDE recovery, Bicchi et al. [1] analyzed aromatic plants and food matrices

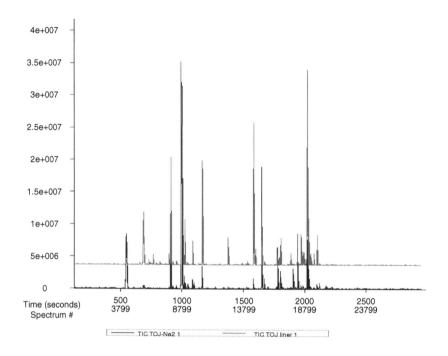

Figure 6.11 Orange juice packaging (top trace) and product (lower trace) using total ion presentation.

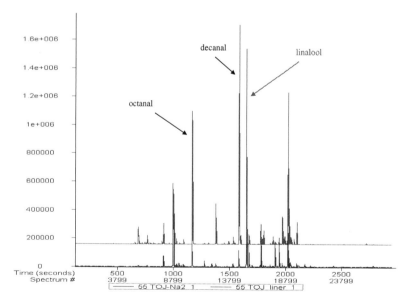

Figure 6.12 SIM presentation using m/e = 55 for important aldehydes and alcohols in juice (lower trace) and liner (upper trace).

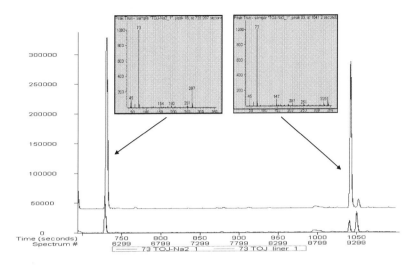

Figure 6.13 Migration of silanes from package (top trace) into juice product (bottom trace).

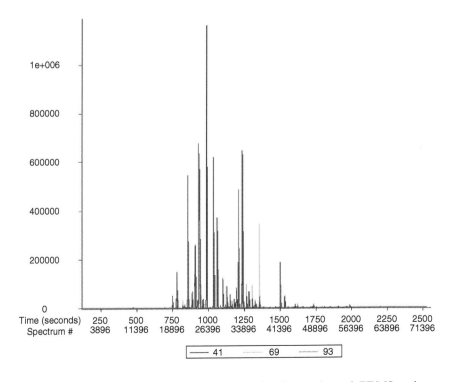

Figure 6.14 SPDE of commercial cola using a 2 ml sample and PDMS carboxen needle. Split ratio: 100:1.

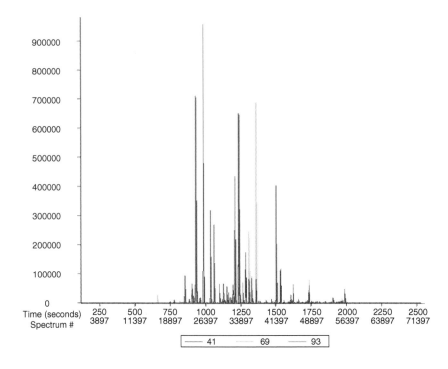

Figure 6.15 SPME of cola using a 2 cm carboxen fiber and identical extraction time as in Figure 2.12. The split ratio is 10:1 whereas the split ratio in Figure 2.5 is 100/1 to accommodate the higher capacity of the SPDE technique.

by HS-SPDE-GC-MS. The study on dried rosemary leaves, green and roasted coffee, white and red wines, and fresh bananas showed that HS-SPDE is a successful technique for HS-sampling with high concentration capability, good repeatability, and reproducibility, also in comparison to HS-SPME. The fresh

Table 6.3 Parameter Changes for SPDE of Cola with Lime Flavor

Experiment	Total	Myrcene	G-Terpinene	Borneol	Cinnamic ald	Myristicin	Function
12 strokes	317,460,000	5,590,300	13,062,000	10,084,000	1,124,700	2,487,700	**Raw area**
at 100 µl/sec		1.76	4.11	3.18	0.35	0.78	**rel%**
25 strokes	326,700,000	6,704,900	14,782,000	12,658,000	1,359,300	3,086,100	**Raw area**
at 100 µl/sec		2.05	4.52	3.87	0.42	0.94	**rel%**
50 strokes	324,110,000	7,440,800	15,160,000	14,442,000	2,063,900	4,562,600	**Raw area**
at 100 µl/sec		2.30	4.68	4.46	0.64	1.41	**rel%**
25 strokes	292,650,000	4,372,900	12,456,000	12,052,000	1,550,200	3,012,300	**Raw area**
at 50 µl/sec		1.49	4.26	4.12	0.53	1.03	**rel%**

Table 6.4 SPME of Cola with Lime Flavor after 10 min Equilibration Time at 50°C (Same Extraction Time as with SPDE)

					Cinnamic		
						SPME	
Experiment	Total	Myrcene	G-Terpinene	Borneol	ald	Myristicin	Function
4.00 min	107,990,000	1,273,700	20,808,000	2,789,000	340,780	725,630	**Raw area**
		1.18	19.27	2.58	0.32	0.67	rel%
8.33 min	86,902,000	1,076,500	16,513,000	2,827,600	622,260	804,100	**Raw area**
		1.24	19.00	3.25	0.72	0.93	rel%
18.66 min	54,853,000	375,460	8,251,500	2,093,000	835,010	928,340	**Raw area**
		0.68	15.04	3.82	1.52	1.69	rel%

banana was equilibrated at 35°C for 15 min, the other samples at 50°C. The following conditions were applied for headspace-sampling by SPDE: agitator (sampling) temperature: 50°C; headspace syringe temperature: 55°C; extraction strokes: 50; plunger speed for extraction: 50 µl/sec; helium volume for desorption: 1 ml; plunger speed for desorption: 15 µl/sec; predesorption time in GC injection port: 30 sec; desorption temperature: 230°C. The trapped analytes were recovered by thermal desorption from the SPDE needle directly into the GC injector body and were analyzed by GC-MS. The parameters were chosen for a good recovery of analytes of medium-to-high volatility and medium-to-low polarity in aqueous media. The low temperature of 50°C favors the adsorption into the polymeric coating, whereas a moderate number of aspiration cycles avoids the loss of most volatile analytes when the plunger draws in the head-space of the next cycle. Dr. Bicchi and his team evaluated the reproducibility on rosemary. The results look similar for SPME and SPDE (Table 6.5).

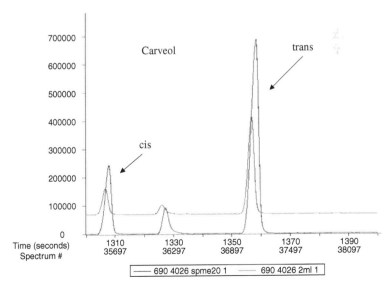

Figure 6.16 Enlargement showing differences between SPDE and SPME in Coke® with lime flavor extraction.

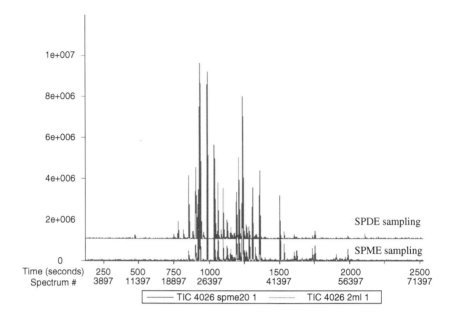

Figure 6.17 Comparison of SPDE with split 100:1 and SPME with split 10:1 of Coke® with lime flavor.

This reports the percent mean peak areas (%) and relative standard deviation, and RSD for a group of 13 compounds identified in the rosemary headspace. HS-SPDE-GC-MS repeatability of the rosemary components was also compared to that of HS-SPME-GC-MS with a PDMS 100-μm fiber applied to the analysis of the same sample.

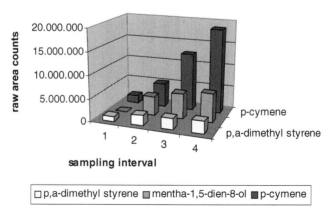

Figure 6.18 Decomposition of citral in acidic media after 0 h, 12 h, 36 h, and 48 h at 30°C.

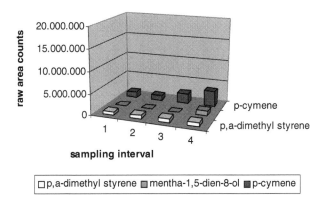

citral protected with stabilizer

□ p,a-dimethyl styrene ■ mentha-1,5-dien-8-ol ■ p-cymene

Figure 6.19 Enhanced stability of citral with additional patented stabilizer.

The RSD, however, was lower in several cases using headspace SPDE compared to headspace SPME. It ranged from 4.1% for camphor to 9.6% for β-ionone. The concentration capability of HS-SPDE was compared to that of HS-SPME. For all analytes investigated in coffee, concentration factors obtained by HS-SPDE were higher than those obtained by HS-SPME. In general, concentration factors increased with the volatility of the compounds and were three to five times higher with SPDE compared to SPME. *p*-Ethyl-guaiacol is an exception, with a lower recovery for SPDE than for SPME, probably because of its low volatility (affects the headspace and matrix

Table 6.5 Comparison of Rosemary by SPDE and SPME

			SPDE			SPME		
	Ret. time		Repeatability		Intermediate			Intermediate
Number	[min]	Compound	Area%	RSD%	precision	Repeatability		precision
1	4.29	α-Pinene	5.2	5.7	6.7	1.8	5.8	6.6
2	8.63	Limonene	4.5	8.7	9.8	1.5	9.1	10.3
3	9.20	1,8-Cineole	2.9	8.2	9.4	1.7	7.7	8.4
4	10.03	Isoamyl alcohol	2.9	7.2	8.6	0.1	10.2	10.8
5	16.62	Linalool oxide	1.1	6.8	7.2	0.7	6.4	7.9
6	18.05	Camphor	20.2	4.1	5.5	12.3	10.2	10.9
7	18.51	3,5-Octadien-2-one	1.6	8.5	8.9	0.8	11.6	11.8
8	19.56	Linalool	4	7	8.4	3.2	10.6	10.9
9	20.03	Bornyl acetate	25.4	6.6	7.1	33.2	11.3	11.6
10	22.64	Verbenone	15.6	9.3	9.6	28	12.4	12.9
11	22.95	Borneol	13.6	8.5	9.8	11.8	9.2	9.9
12	27.79	β-Ionone	0.4	9.6	9.7	1.5	7.1	8.1
13	32.68	Thymol	2.8	7.2	7.8	3.5	4.1	5.3

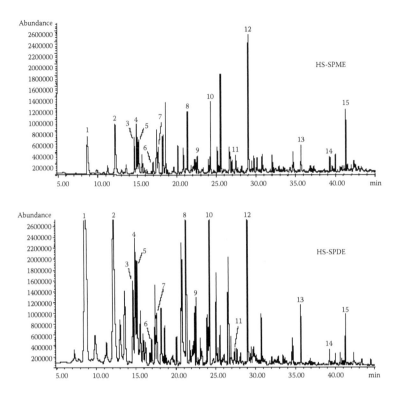

Figure 6.20 GC-MS profiles of roasted coffee after HS-SPME and HS-SPDE samplings.

partition) and its polarity (affects its solubility in PDMS). The difference in concentration factors can be explained by the fact that the volume of PDMS coating the SPDE needle wall is about eight times higher than the one coating the SPME fused silica fiber. Thanks to this fact, HS-SPDE achieves a high concentration capability in particular for high-volatility analytes (V_{SPDE}: 4.5 µl vs. V_{SPME}: 0.6 µl), are shown in Figure 6.20.

E. Amitraz in honey

In Germany, honey is regularly tested for amitraz, a pesticide used against mites on fruit trees. The food safety regulatory office permits a level of amitraz in honey of 200 µg/kg. Until 2003, a costly analytical determination was being applied. It first used the hydrolyzed alkenes of amitraz. The newly formed 2,4-dimethyl aniline (see Figure 6.21) was isolated through the Clevinger distillation. Then, however, headspace-SPDE was introduced by H. Hahn et al. [4]. Only concentrations near the limit have to be confirmed with the officially approved method. 4-Chloro-2-methyl aniline is used as an internal standard.

A calculated limit of detection (LOD) of 140 µg/kg and a limit of quantitation (LOQ) of 200 µg/kg was determined for amitraz in honey. Mass

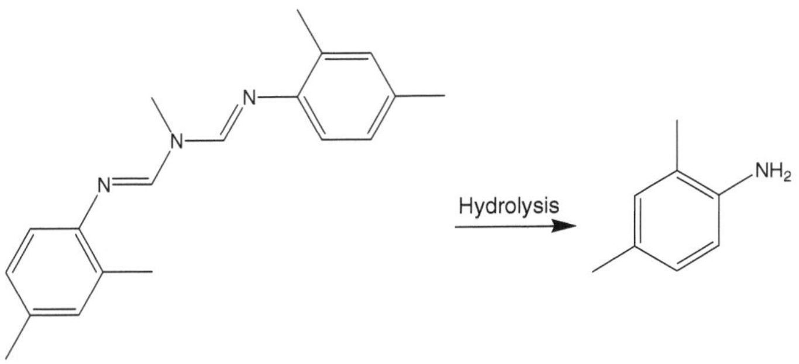

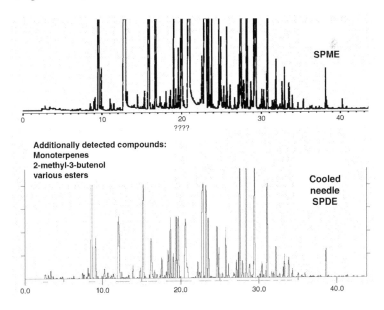

Figure 6.21 Hydrolysis of amitraz to 2,4-dimethyl aniline.

spectrometric identification is possible down to a concentration of 15 µg/kg. The recovery is about 87 to 99%.

F.   Analysis of highly volatile compounds with SPDE extraction

The ambient temperature of the SPDE needle sometimes discriminates highly volatile compounds. This limitation is also well known for SPME. To trap highly volatile compounds more effectively, the Extraction Cooler cools the needle down to 15°C.

We applied cooled needle SPDE as well as SPME to citrus samples like orange oil and orange essence oil. According to their findings, monoterpenes, 2-methyl-3-butenol and various esters could be detected with the new cooled needle SPDE. Due to their low boiling point, they cannot be separated with SPME (Figure 6.22).

Figure 6.22 Comparison of SPME and cooled needle SPDE on orange oil.

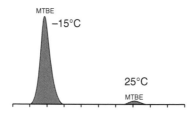

Figure 6.23 Side-by-side comparison of MTBE at ambient temperature and −15°C.

The extraction cooler was also used for the analysis of MTBE by CHROMTECH GmbH. The chromatogram (see Figure 6.23) shows the concentration difference at ambient compared to the triple-peltier cooled temperature.

III. Conclusions

SPDE is a superior technique when automation and capacity of other techniques is not available. SPDE has proven to be a very sensitive method for the analysis of volatiles in beverages and other food matrices. The high volume of polymer coating is especially useful for the concentration of highly volatile analytes, because other techniques often fail in that area. Customized polymers used as stationary phases may give way to new possible extraction pathways and extraction possibilities. One other advantage is the durability of the SPDE needle, which allows to complete long-term studies using the same syringe, without the necessity to verify that the previous device was really similar.

SPME will remain an important technique and may even be the first technique attempted with unknown samples to determine if headspace sampling is a viable option. If successful, the analysis can be moved to SPDE for increased sensitivity and a higher degree of automation and longer lifetime of needles.

References

1. Carlo Bicchi, Chiara Cordero, Erica Liberto, Patrizia Rubiolo, and Barbara Sgorbini, Automated headspace solid-phase dynamic extraction to analyse the volatile fraction of food matrices, *Journal of Chromatography A*, 1024, 217–226, 2004.
2. Louis Trauth Dairy, Inc. — Grade A Pasteurized — Fat Free Skim Milk — Vitamin A & D (best before Oct 31D).
3. Tropicana — Pure Premium — Not From Concentrate — ORIGINAL — NO PULP — 100% Pure Squeezed Orange Juice — 8 FL OZ (240 ml) — PASTEURIZED — (best before NOV 25 48GL2034).
4. Chemisches und Veterinaeruntersuchungsamt Sigmaringen in 2001 for screening for amitraz (described in nach § 35 LMBG L 00.00-58: Gaschromatographische Bestimmung von Amitraz und Vinclozolin sowie ihren 2,4-Dimethylanilin bzw. 3,5-Dichloranilin enthaltenden Metaboliten in Lebensmitteln, Stand Juli 2000, Beuth-Verlag — German law of consumer protection in farmed products).

chapter seven

The application of chemometrics for studying flavor and off-flavor problems in foods and beverages

Ray T. Marsili

Contents

I. Introduction

It is no simple task to decipher which chemicals in a complicated chromato-gram of dozens or hundreds of chemical peaks, as in the case of coffee samples, are responsible for the flavor characteristics of a sample being stud-ied. As indicated in examples throughout this book, olfactometry, odor-unit calculations, model systems studies, and numerous other techniques have been used by flavor researchers to aid in this process. Another tool is chemo-metrics. This chapter provides more details on the application of chemometrics/multivariate analysis (MVA) to flavor research, as well as its application potential as a rapid screening tool for flavor quality-control applications.

What is chemometrics? [1] Data collection in science usually involves many measurements performed on numerous samples. Such multivariate data is commonly analyzed using one or two variables at a time. Unfortu-nately, this approach usually fails to discover subtle but important relation-ships among all samples and variables. To correct this problem, all data must be processed/analyzed simultaneously. This is the purpose of chemometrics.

Typically, chemometrics is used for three primary purposes:

- To explore patterns of association in data
- To track properties of materials on a continuous basis
- To prepare and use multivariate classification models

Exploratory data analysis: Patterns of association exist in many data sets, but the relationships between samples can be difficult to discover when the data matrix exceeds three or more features. Exploratory data analysis can reveal hidden patterns in complex data by reducing the information to a more comprehensible format. Such a chemometric analysis can expose possible outliers and indicate whether there are patterns or trends in the data. Exploratory algorithms such as principal component analysis (PCA) and hierarchical cluster analysis (HCA) are designed to reduce large complex data sets into a series of optimized and interpretable views. These views emphasize the natural groupings in the data and show which variables most strongly influence those patterns.

Continuous property regression: In many applications, it can be extremely challenging to measure a property of interest directly. Such cases require the analyst to predict something of interest, based on related prop-erties that are easier to measure. The goal of chemometric regression analysis is to develop a calibration model which correlates the information in the set of known measurements to the desired property. Chemometric algorithms for performing regression include partial least squares (PLS) and principal component regression (PCR), and are designed to avoid problems associated with noise and correlations in the data. Because the regression algorithms used are based in factor analysis, the entire group of known measurements is considered simultaneously, and information about correlations among the variables is automatically built into the calibration model. Chemometric regression lends itself handily to the online monitoring and process-control industry, where fast and inexpensive systems are needed to test, predict, and make decisions about product quality.

Classification modeling: Many applications require that samples be assigned to predefined categories, or "classes." This may involve determining whether a sample is good or bad, or predicting an unknown sample as belonging to one of several distinct groups. A classification model is used to predict a sample's class by comparing the sample to a previously analyzed experience set, in which categories are already known. K-Nearest Neighbor (KNN) and Soft Independent Modeling of Class Analogy (SIMCA) are primary chemometric workhorses. When these techniques are used to create a classification model, the answers provided are more reliable and include the ability to reveal unusual samples in the data. In this manner, a chemometric system can be built that is objective and thereby standardizes the data evaluation process.

II. Multivariate analysis approaches for flavor studies

There are numerous ways in which the chemometric algorithms of exploratory analysis, classification modeling, and continuous property regression can be applied to flavor applications. One way of appreciating the power of chemometrics is to consider its application potential in the food industry (1) as a rapid QC tool for flavor profiling finished products and raw materials (e.g., as acceptable or unacceptable in flavor) and (2) as a research tool for uncovering the chemicals that are most significantly impacting a product's characteristic flavor profile.

Figure 7.1 shows a conventional GC-MS chromatogram of a milk sample contaminated with 1300 ppm Matrixx, a sanitizer used in milk processing lines after cleaning. Milk contaminated with Matrixx has an oxidized-type

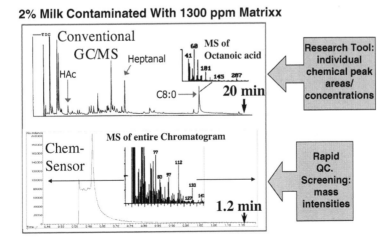

Figure 7.1 Conventional GC-MS vs. ChemSensor®.

flavor, which is difficult to distinguish by taste from light-abused milk. The active ingredients in Matrixx are peroxyacetic acid and octanoic acid. Oleic acid in milk can be oxidized to heptanal by hydrogen peroxide in Matrixx. The presence of three key chemicals — octanoic acid, acetic acid (a decomposition product of peroxyacetic acid), and heptanal — is a good indication that a potential off-flavor problem from sanitizer contamination is possible. Whereas conventional GC-MS testing can be used to confirm Matrixx contamination of processed milk, the problem with this approach is chromatographic time (at least 20 min), and subsequent chromatogram interpretation are prohibitively time-consuming if dozens of milk samples need to be tested at one time.

A more practical approach for efficient screening of a large number of milk samples for sanitizer contamination is based on the application of the ChemSensor® (Gerstel Inc., Baltimore, MD). In this approach, the analytical column (normally 30 m long) is replaced by a retention gap, an uncoated fused-silica capillary column approximately 1 m in length. Analysis time is reduced from 20 min to less than 2 min. What is sacrificed, however, is the ability to determine the presence of individual chemical peaks. By analyzing numerous control (normal-tasting) milk samples and several samples intentionally contaminated with sanitizer, multivariate analysis techniques can then be applied to "train" the chemometrics software to distinguish mass intensity patterns that correspond to Matrixx-contaminated samples from mass intensity patterns that correspond to control samples. Chemometrics software can be used to develop a classification model; once created, the model can then be used to make class predictions (i.e., sanitizer-contaminated or not contaminated by sanitizer) of unknown milk samples. Once properly calibrated, the entire process from sample testing to sample classification is completed automatically at a rate of one sample every 2 to 3 min. Several articles have been written illustrating examples of the use of the Gerstel Chemsensor as a rapid QC screening tool for food flavor applications [2–7].

There is another way chemometrics can be applied to the Matrixx contamination problem. If the markers/byproducts (octanoic acid, acetic acid, and heptanal) of Matrixx contamination were unknown, it should be possible to identify these chemicals with the support of multivariate analysis. In this case, several control milk samples and several samples of Matrixx-contaminated milk could be analyzed by conventional GC-MS. In this case, the independent variables plugged into the chemometric spreadsheet would be peak areas of specific chemicals appearing in the chromatograms (rather than mass intensities of specific masses, as in the case for the ChemSensor approach). The PCA algorithm can be applied to this type of data. The goal would be to show a cluster for control samples well-separated from a cluster of Matrixx-contaminated samples. Loadings plots and tables could then be examined to determine which specific chemicals were driving the clustering/ modeling of the two types of sample classes. This approach can prove quite powerful for elucidating chemicals responsible for off-flavors. It is more time-consuming than ChemSensor analysis but provides more detailed results regarding the flavor influence of specific chemicals in samples. This is an especially powerful technique when there are only subtle changes in chromatograms of control and abused-sample classes. PCA studies are often capable of finding meaningful chemical differences between classes of samples — even differences that are impossible to find by visual inspection of raw data files.

Following are more examples explaining how chemometrics can be used for resolving flavor problems.

A. MS E-Nose applications based on mass intensity data

1. Classification of coffee samples by geographic origin

Figure 7.2 shows classification clusters of four different coffee types (Guatemalan, decaffeinated Guatemalan, Sumatran, and decaffeinated Sumatran) generated from static headspace GC-MS testing using the Gerstel ChemSensor. The independent variables considered were mass intensities for 120 different atomic mass units (m/z 51 to 170). Inspection of Figure 7.2 shows one of the decaffeinated Guatemalan samples is potentially an outlier. A two-dimensional plot of sample residuals vs. Mahalanobis distances provides a good indication of samples that are possible outliers and should be excluded from data analysis (Figure 7.3). Figure 7.4 shows a two-dimensional PCA plot of the four classes of coffee samples after the decaffeinated Guatemalan outlier was excluded from the data set.

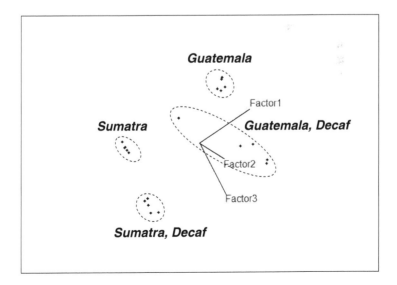

Figure 7.2 3-D PCA plot of four classes of coffees (five samples for each class) based on ChemSensor® static headspace testing using mass intensity results from 51 to 170 amu.

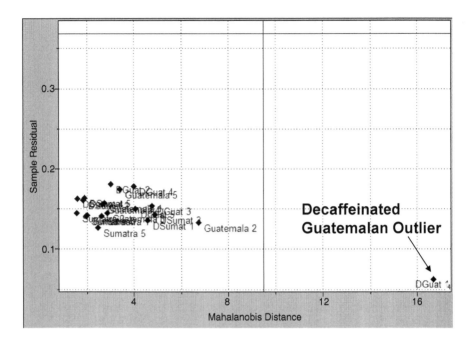

Figure 7.3 Determination of decaffeinated Guatelmalan outlier sample from Mahalanobis Outlier Diagnostics.

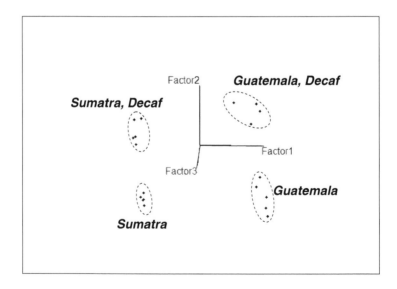

Figure 7.4 3-D PCA plot of four classes of coffees based on ChemSensor® static headspace testing using mass intensity results from 51 to 170 amu after deletion of decaffeinated Guatemalan outlier.

With excellent clustering of similar sample types, these results could be used to create a KNN model for class prediction. The model could then be used to analyze coffee samples of unknown origin in order to determine their proper geographic class origin.

This example illustrates the diagnostic power of chemometric software (Pirouette® from Infometrix, Bothell, WA) to determine outliers. It also shows how "loadings" plots can be used to determine the independent variables most influential in clustering of sample classes (Figure 7.5). In this case, the loadings plots for Factor 1 show that mass intensities for m/z 52, m/z 60, and m/z 79 were most influential, and for Factor 2 the most important masses for clustering were the same three masses plus m/z 81, m/z 95, and m/z 98. While more research is probably needed, these results could indicate that acetic acid (m/z 60) and pyridine (m/z 79), which are known to be important flavor contributors to some coffees, are most likely strongly responsible for differentiating classes for these four types of coffees.

Note that if this experiment was done in a slightly different way, using conventional GC-MS and measuring the peak areas of specific peaks rather than using the ChemSensor approach, it would likely be possible to identify the chemical constituents in these coffee samples that define their specific classes. Whereas this approach would be more time-consuming, the research insights that could be revealed might be well worth the effort.

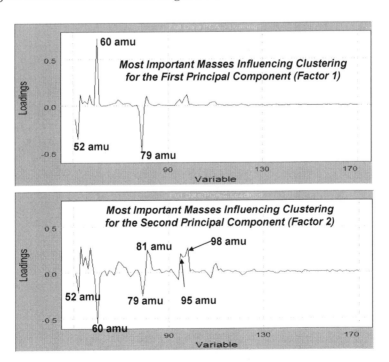

Figure 7.5 Examination of PCA loadings plots for coffee samples to determine masses most influential to clustering.

2. Understanding and predicting the cause of milk off-flavor using MVA techniques

Milk is susceptible to developing off-flavors through myriad pathways [8]. Milk is an excellent medium for the growth of microorganisms. Malodorous chemicals produced as metabolites during the bacteria's growth phase are a common cause of acid, malty, fruity, bitter, putrid, and unclean off-flavors in dairy products.

Postpasteurization contamination of milk with psychrotrophic bacteria can cause milk off-flavors. Some characteristic off-flavor notes in milk are clearly associated with specific types of psychrotrophs. For example, contamination by *Pseudomonas fragi*, a psychrotrophic Gram-negative organism, often causes "fruity" off-notes in milk. *Pseudomonas fragi*'s lipase and esterase enzymes hydrolyze short-chain fatty acids from milk fat and convert the acids to ethyl esters by reaction with ethanol. Strains of *Bacillus* spp. have also been observed to produce fruity off-flavors in milk.

Another common psychrotroph associated with a particular flavor defect is *Streptococcus lactics* var. *maltigenes*. The malty aroma of milk cultures of *Streptococcus lactics* var. *maltigenes* is caused by the production of 3-methylbutanal. 3-Methylbutanal concentrations as low as 0.5 ppm in milk generate the characteristic malty flavor defect. 3-Methylbutanal is produced from the action of bacterial enzymes on leucine. Another organism capable of producing 3-methylbutanal is *Lactobacillus maltaromicus*.

Commonly, the shelf life of processed milk — the period between processing and the time when milk becomes unacceptable to consumers because of taste or odor — depends on the numbers and types of organisms that contaminate milk after pasteurization. Typically, dairies attempt to estimate the shelf life of newly produced products with the Mosely Keeping Quality test or similar types of microbiological-based tests. Microbiological tests do not normally correlate very well to actual shelf life, probably because these tests measure total microbial counts, which reveal nothing about the types and levels of off-flavor metabolites being generated.

a.   Partial least squares (PLS) analysis of processed milk to predict shelf life. A method that can better predict the shelf life of freshly processed milk was developed based on an MS e-nose approach using SPME with a Carboxen-PDMS fiber followed by GC-MS on a 1 m uncoated fused silica column instead of the usual 30 m long fused-silica capillary column [9]. Three milliliters of milk sample, 5 µl of internal standard solution (10 µg/ml chlorobenzene) and a micro-stirring bar were placed in a 6 ml glass GC vial and capped with 20 mm PTFE/silicone septa.

The mass range used was m/z 50 to m/z 150. A mass intensity list was obtained for each sample by averaging the masses between 100 sec and 500 sec. The mass intensities were then normalized by dividing by the intensity of the major mass peak for the chlorobenzene internal standard (m/z 112). For PLS calculations, these normalized mass ratios from m/z 50 to m/z 150 were used as independent variables, and the shelf life determined by sensory analysis was used as the dependent variable. Details of the procedure have been previously reported [9].

All samples consisted of commercially pasteurized and homogenized reduced-fat milk (2% milk fat), free of off-flavors at time of manufacture. Samples were packaged in either pint or half-pint high-density polyethylene (HDPE) contoured bottles with screw caps. Thirty samples of reduced-fat milk were sampled consecutively from the production line at a dairy plant on the day of processing. This sampling scheme was conducted on six occasions over a 7-month period.

Samples were immediately taken from the dairy plant and refrigerated in a walk-in cooler at 7.2 ± 0.5°C until the end of shelf life. During refrigerated storage, two bottles of reduced-fat milk were removed for testing at predetermined intervals — three times weekly in the initial stage of refrigerated storage, and then daily when a decline in flavor quality was observed. One sample from each pair was subjected to organoleptic evaluation, and one sample was placed in a 19 ± 1°C incubator for 16 h. After 16 h, the preincubated sample was subjected to SPME-MS-MVA analysis.

Shelf life prediction with the ChemSensor-based test approach is not only more accurate than traditional microbiological-based shelf life prediction methods, but it is also faster.

Table 7.1 compares actual shelf life (determined by sensory evaluation) to predicted shelf life for the 20 samples of reduced-fat milk that were not used to make the PLS model. The 20 samples were tested blind; i.e., their shelf life was not known at the time of testing. On average, the

Table 7.1 Actual[a] and Predicted[b] Shelf Life (in Days) of Pasteurized and Homogenized Reduced-Fat Milk

Actual	Predicted	Error[c]	Actual	Predicted	Error[d]
15	15.3	0.3	4	5.0	1.0
18	18.2	0.2	3	4.2	1.2
14	14.7	0.7	2	2.3	0.3
8	7.4	−0.6	0	0.7	0.7
5	5.4	0.4	10	12.8	2.8
1	1.8	0.8	7	7.3	0.3
14	13.6	−0.4	3	3.5	0.5
11	10.3	−0.7	2	2.6	0.6
10	10.2	0.2	5	4.9	−0.1
7	6.4	−0.6	0	0.0	0.0

Note: Predictions based on mass intensity data for masses 50 to 149, excluding masses at m/z 59, m/z 73, and m/z 77). Sixty-four samples used to develop prediction model. The 20 samples below, which were not used to make the PLS model, were analyzed to see how closely the predicted shelf life value matched the actual value. Error range (days) = −0.7 to +2.8.; R^2 = 0.9801.

[a] Determined by sensory panel.

[b] Predicted from SPME-MS mass intensity data using PLS prediction models.

[c] Error = Predicted − Actual.

[d] Average error (days) = ±0.62.

Source: R.T. Marsili, *J. Agric. Food Chem.* 48: 3470–3475, 2000. With permission.

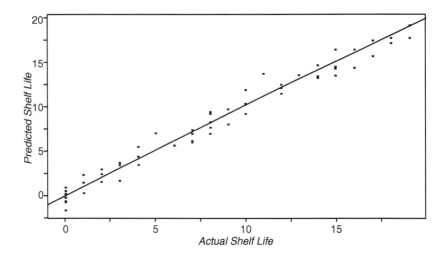

Figure 7.6 PLS plot of predicted shelf life (based on SPME MS mass intensity data) vs. actual shelf life (based on sensory testing) for the 64 samples used to prepare the PLS model for reduced-fat milk. (From R.T. Marsili, *J. Agric. Food Chem.* 48: 3470–3475, 2000. With permission.)

SPME-MS-MVA PLS model for reduced-fat milk predicted the shelf life with an accuracy of ±0.62 d, with a correlation coefficient of 0.9801 and a range of –0.7 to +2.8 d. Figure 7.6 shows a plot of predicted vs. actual shelf life (based on sensory testing) for the 64 samples used to prepare the PLS model for reduced-fat milk.

    b.   Determining the cause of off-flavor in processed milk.   Another experiment was conducted to see what type of shelf life prediction would be made for samples with off-flavors derived from nonmicrobiological causes and if principal component analysis (PCA) plots could be applied as a tool to help identify two relatively common nonmicrobial causes of off-flavor. Using PCA plots of SPME-MS-MVA data and modeling with KNN and SIMCA to classify milk samples as to the cause of off-flavor has been previously reported [10,11].

    Reduced-fat milk samples were spiked (aseptically with a 5 ml glass syringe through a pin hole in the top of the cap) with either copper (C) or Matrixx sanitizer (S), incubated at 19°C for 16 h, and then analyzed by SPME-MS-MVA to generate normalized mass-intensity data as was done for the PLS shelf life prediction study. Fresh reduced-fat milk (F) was used as a control, and bacteria-abused milks (B) were obtained by sampling unopened bottles with 0 d of shelf life. The 16 h incubation period allowed time for microbes to grow and produce detectable levels of metabolites and also allowed time for copper and sanitizer to produce their unique types of off-flavor chemicals. All samples were packaged in half-pint HDPE bottles and were from the same day's production.

The "C" samples were spiked with copper sulfate at a concentration of 5 ppm copper in milk, and the "S" samples were spiked with Matrixx at a concentration of 1,300 ppm in milk. These levels of abuse agents were selected because sensory testing determined that this was the threshold taste level for these off-flavor causing agents. Reduced-fat milk with 5 ppm copper was judged organoleptically to have a shelf life of 1 d. The milk spiked with 1,300 ppm Matrixx was judged to have a shelf life of 0 d. The actual and predicted shelf life of these samples appear in Table 7.2, and a PCA two-dimensional plot of SPME-MS-MVA data appears in Figure 7.7.

This experiment showed that whereas SPME-MS-MVA was not able to accurately predict shelf life for copper- and sanitizer-abused samples, the technique does provide strong indication that there is a shelf life problem for these samples. This suggests the shelf life prediction technique could be refined by one of the following two approaches:

1. To improve prediction accuracy for copper- and sanitizer-abused samples, a number of these types of abused samples could be included

Table 7.2 Actual (Determined from Sensory Panel) and Predicted[a] Shelf Life of Fresh (F), Sanitizer-Contaminated (S), Copper-Contaminated (C), and Bacteria Spoiled[b] (B) Pasteurized and Homogenized Reduced-Fat Milk

Sample	Actual shelf life (Days)	Predicted shelf life (days)	Error[c] (days)
F1	20	19.7	−0.3
F2	20	20.1	0.1
F3	20	22.0	2.0
F4	20	18.6	−1.4
F5	20	18.2	−1.8
S1	0	10.6	10.6
S2	0	0.1	0.1
S3	0	6.4	6.4
S4	0	0.0	0.0
S5	0	6.3	6.3
C1	1	−2.4	−1.4
C2	1	8.9	7.9
C3	1	9.4	8.4
C4	1	11.3	10.3
C5	1	12.7	11.7
B1	0	0.2	0.2
B2	0	0.0	0.0
B3	0	0.6	0.6
B4	0	0.1	0.1
B5	0	0.5	0.5

Note: Predictions based on mass intensity data for masses 50 to 149, excluding masses at *m/z* 59, *m/z* 73, and *m/z* 77.

[a] Predicted from SPME-MS-MVA data using the same PLS prediction model as was used for reduced-fat milk in Table 7.1.

[b] Tested at the end of shelf life (after 20 d of incubation at 7.2°C).

[c] Error = Predicted − Actual.

Source: R.T. Marsili, *J. Agric. Food Chem.* 48: 3470–3475, 2000. With permission.

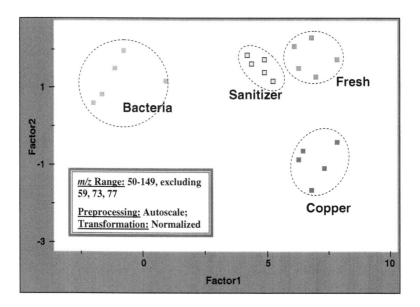

Figure 7.7 2-D PCA discrimination between different causes of milk off-flavor.

in future PLS modeling experiments. This more comprehensive PLS model could then be applied to freshly processed milk samples to determine their potential for untimely spoilage (i.e., off-flavor development caused by bacterial metabolites, lipid oxidation, or sanitizer contamination).

2. A KNN or SIMCA prediction model could be created to not only include bacterial spoiled samples but also samples spoiled by copper-based oxidation, sanitizer abuse, and even other commonly encountered causes of milk off-flavor development (e.g., photo-oxidized milk). Finished milk products could then be rapidly tested using the ChemoSensor approach to determine potential nonmicrobial off-flavor problems before the product is shipped to the marketplace.

B. Application of chemometrics for understanding the role of specific chemicals that define characteristic food and beverage flavors

In some cases, rapid screening of various classes of foods or raw materials or prediction of shelf life, flavor score, or some other continuous property are not the primary goals in conducting flavor experiments. Often, instead, food scientists want to know more details about specific analytes that influence flavor. The application of chemometrics using peak areas/concentrations of analytes in place of mass intensities, as described previously in this chapter, is a valuable tool for this type of problem. While more information is gained about the influence of specific analytes on the problem being studied, the disadvantage encountered is considerably longer analysis time because of the need to attain sufficient peak resolution for accurate quantitation. As a result, this application of chemometrics is a research tool rather than a rapid QC screening tool. Often, however, the tradeoff is worth it. Following are two examples.

1. Determination of the age of sherry wines by chemometrics

Wines are consumed after aging in wooden casks, in the bottle, or in both, successively. Aging modifies the sensory properties of the wine. Aging is an expensive process for the winemaker, and wine consumers pay premium prices for the resulting flavor. Among the chemicals extracted from wood casks are vanillin and other aldehydes, benzoic acids (particularly gallic acid), and furan derivatives.

A better understanding of the chemical reactions that take place during wine aging might enable winemakers to modify the final flavor of the product, enhancing desirable characteristics, and reducing undesirable characteristics of the final wine.

Correlating the concentration of various compounds in wine with the wine's storage period in wood casks has been challenging. The concentration of vanillin in many wines, for example, far exceeds quantities that could possibly be extracted from the wood during aging. It is difficult to correlate furan compounds, which are extracted in relatively large quantities from the wood, with aging because these compounds are authorized for use as wine additives. Another problem encountered with getting accurate predictions of aging from chemical profiling is that some compounds extracted from wood increase in concentration in the wine initially during storage in wood and then experience concentration decreases with further aging, probably because of reactions with other compounds present in wine.

As it has not been a simple task to correlate the concentration of specific compounds with the age of wine, researchers have turned to chemometric techniques. Dominico Guillen and coresearchers [12] recently were able to determine the age of Sherry wines by regression techniques using routine analytical testing parameters, as well as concentrations of phenolic and volatile compounds. One goal of this research was to develop a regression-prediction model that could guarantee the age of wine from objective chemical measurements. A second objective was to elucidate the chemical phenomenon that occurs during wine aging.

Guillen's study involved a total of 30 wine samples of known ages. The dates of production ranged from 1932 to 1999. The wines were stored under similar conditions and were of the same variety. In addition, the method of production employed for all wine samples was the same. A second group of five samples, also vintage wines but whose ages were unknown, was analyzed and subjected to regression analysis to test the accuracy of the regression predictions. This second group of five samples was produced and aged the same way as the first group of 30 samples.

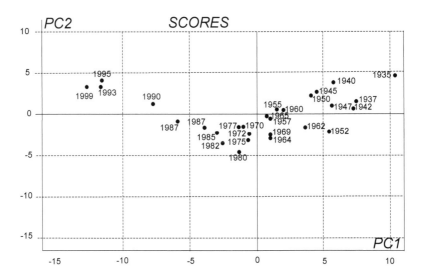

Figure 7.8 PC1 vs. PC2 scores plot of wine samples obtained in the PLS calibration model. (From D.A. Guillen, M. Palma, R. Natera, R. Romero, and C.G. Barroso, *J. Agric. Food Chem.* 53: 2412–2417, 2005. With permission.)

Phenolic compounds were quantitated by HPLC. Quantitation of volatiles was measured by injection of 1 µl of wine distillate containing 4-methyl-2-pentanol internal standard. Regression techniques used included multiple linear regression (MLR) and partial least squares regression (PLS).

PLS results for this study are shown in Figure 7.8 and Figure 7.9. The PCA scores plot for the first two principal components (PCs) is shown in Figure 7.8.

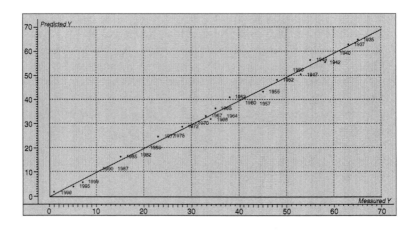

Figure 7.9 Regression line for plot of actual vs. predicted age of sherry wines using the PLS calibration model. (From D.A. Guillen, M. Palma, R. Natera, R. Romero, and C.G. Barroso, *J. Agric. Food Chem.* 53: 2412–2417, 2005. With permission.)

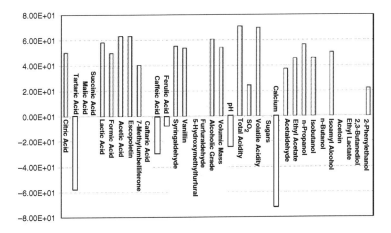

Figure 7.10 Contribution of the individual variables to the first principal component (PC1) of the PLS regression model. (From D.A. Guillen, M. Palma, R. Natera, R. Romero, and C.G. Barroso, *J. Agric. Food Chem.* 53: 2412–2417, 2005. With permission.)

Data points show a good spread (i.e., lack of clustering), as would be expected in a well-behaved PLS model. The regression line drawn against the prediction obtained in the PLS calibration model is shown in Figure 7.9.

Figure 7.10 shows the contribution of each variable to PC1. Notable findings are the negative contributions of tartaric acid and calcium (probably because of a precipitation mechanism), the positive contribution of the acids related to oxidative aging and acetic and formic acids, and the total and volatile acidity, together with the positive contributions of phenolic compounds related to extraction from wood (escopoletin, 7-methylumbelliferone, and vanillin).

PLS and MLR prediction accuracies for five unknown wine samples are shown in Table 7.3. As the PLS model provided more accurate predictions than the MLR model, the MLR calibration and results are not discussed in detail here.

This study showed that PLS and MLR statistical regression methods — utilizing normal routine analytical parameters for wine QC testing, polyphenols,

Table 7.3 Predicted Age for Five Unknown Vintage Sherry Wines by PLS and MLR

	PLS		MLR	
Sample	Age prediction (Years old)	Deviation	Age prediction (Years old)	Deviation
Clotilde	20.249	± 3.233	23.340	± 6.880
Cristeta	31.439	± 2.627	30.390	± 4.005
Lajulia	30.953	± 5.835	38.700	± 10.383
Olorosodi	30.435	± 3.897	33.095	± 2.207
Solerae	22.380	± 11.425	16.606	± 10.506

Source: D.A. Guillen, M. Palma, R. Natera, R. Romero, and C.G. Barroso, *J. Agric. Food Chem.* 53: 2412–2417, 2005. With permission.

organic acids and phenolic compounds as predictive variables — can be used to determine the age of a vintage Sherry wine. It also showed that PLS, a tool used extensively in chemometrics, is more accurate for predicting wine age than MLR.

Several papers on the use of chemometrics for studying wine, spirits, and beer flavor issues have been published in recent years [13–20].

2. Determining chemicals responsible for off-flavor in club cheese

A producer of cheese-powder ingredients was having a problem with abnormally high rejections of cheese powders, because the product failed to pass sensory testing. The product was often criticized as being too sour, musty, or dirty-tasting. A considerable dollar loss was involved with the rejections.

The cheese-powder producer traced the problem to a club cheese ingredient used in the formulation. The customer of the cheese-powder supplier insisted that this particular type of club cheese be used in its product formulations because it produced a desirable earthy, mushroom-like flavor in the finished cheese powder. This flavor nuance was highly desirable and difficult to achieve with other types of club cheeses or other cheese ingredients. The cheese powder producer, on the other hand, was reluctant to use the recommended club cheese because (1) it was the likely source of the product rejection problem, (2) it was often received with considerable surface-mold growth and had to be returned, and (3) it was unusually expensive.

The ingredient supplier conducted an analytical study to learn the answers to the following questions:

- What were the chemicals responsible for the desirable flavor attributes of the club cheese?
- What were the chemicals responsible for the undesirable sour, musty, dirty flavor notes that sometimes caused product rejection?
- How could the rejection problem be eliminated (or at least drastically reduced)?

The chemists attempted to find answers to as many of these questions as possible. Indeed, not only did they answer all the questions, but they also learned even more interesting facts about cheese flavor that could be helpful for future problems. Subjecting peak-area data from club cheese samples tested by SPME GC-MS to PCA and examination of loadings and modeling-power results were the primary tools for determining the key flavor chemicals in the club cheese.

Thirteen club cheese samples were analyzed: seven control (good, normal-tasting), three borderline (flavor not as good as control but not as bad as rejects), and three reject samples (samples that did not pass sensory testing). The analytical technique used was SPME GC-MS [21]. One gram samples of cheese powder plus 5 ml of 25% NaH_2PO_4 solution were preincubated 4 min at 50°C and then extracted with a PDMS/DVB fiber for 20 min at 50°C. Forty-eight organic volatiles were measured by this technique.

Unsupervised pattern recognition proved to be a useful approach for visualizing important relationships in data. Two-dimensional PCA plots showed good clustering of the three classes of club cheese (Figure 7.11). The excellent clustering demonstrated for the three sample classes showed that the chemicals responsible for modeling the flavor characteristics of these samples were being extracted by the analytical method employed. Therefore, there was no need to explore alternate sample-preparation methods.

Inspection of loadings graph and table and the modeling power table (Table 7.4) showed that known mold metabolites were important independent variables for modeling. Four of these chemicals (those shown in bold in the table) are known mold metabolites. This would indicate that mold development played a major role in the modeling of these samples. The club cheese was sold as a cheddar-flavored processed cheese — not one made from mold-ripened cheese.

After only four days in the refrigerator, most of these samples developed excessive black surface mold. This mold was scraped from samples and analyzed in triplicate using the identical SPME GC-MS method used to analyze the original samples. Table 7.5 lists the volatiles at the highest levels in the mold samples scraped from the surface of the cheese samples. Several of these chemicals were significant in modeling the three classes of club cheese, further substantiating that mold metabolites were strongly influencing flavor-acceptability of the club cheese.

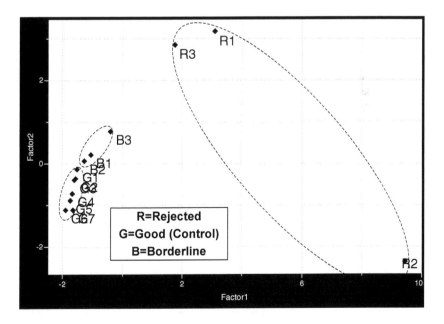

Figure 7.11 2-D PCA plot of rejected, good, and borderline samples of club cheese based on SPME GC-MS peak area data of 48 organic volatiles. (Note: classifications of rejected, good, and borderline based on sensory testing by trained flavor panel.)

Table 7.4 Volatiles in Club Cheese with Highest Modeling Power

Rank	Chemical	Modeling power
1	Isopropyl octanoate	0.968
2	**2-Undecanone**	0.958
3	**8-Nonen-2-one**	0.957
4	**2-Nonanone**	0.956
5	2-Propanol	0.943
6	Nonanal	0.943
7	Amyl acetate	0.940
8	**2-Heptanone**	0.923
9	2-Heptanol	0.921
10	2-Methyl-1-butanol	0.915

Note: Chemicals indicated in bold are known mold metabolites observed in mold-ripened cheese.

As a result of these findings, the following conclusions were made:

• The unusual appealing flavor of this particular type of club cheese is due to mold metabolites and a mite pheromone (8-nonen-2-one).
• Rejection of some club cheese samples and powders made from certain lots of the club cheese was caused by excessive amounts of mold growth and mold metabolites.
• The supplier needs to maintain tighter QC controls for mold growth (limit rework).
• QC monitoring of mold growth and off-flavors could readily be accomplished with the Gerstel ChemSensor. Development of an accurate and rapid ChemSensor screening test for excessive mold growth for this particular club cheese is highly likely because a suitable analytical method was already developed (SPME GC-MS) and the specific chemical markers (and their major mass peaks) were clearly identified.

Table 7.5 Volatiles Present at High Levels in Surface Scrapings of Moldy Club Cheese Samples

Chemical	Flavor descriptor
p-Methyl anisole	Dirty, musty
1-Octen-3-ol	Mushroom
2-Heptanol[a]	Mushroom-like/sour
8-Nonen-2-one[a,b]	Fatty, earthy
Isobutyl alcohol	Vinous
Isoamyl alcohol	Vinous
1-Pentanol	Vinous
Isopropyl octanonate[a]	Fruity
Amyl acetate[a]	Sweetish
2- and 3-Octanone	Soapy and resinous

[a] High modeling power in PCA plots.
[b] Significant flavor component of blue and mold-ripened cheese.

The club cheese producer, anxious not to lose the business from the cheese powder manufacturer, drastically reduced its common practice of excessive rework for the club cheese and, as a result, flavor problems with the club cheese and the cheese powder made with the club cheese were almost completely eliminated.

One interesting additional point learned about these flavor chemicals is that 8-nonen-2-one, which has a unique and desirable piquant flavor, is not actually a mold metabolite *per se*. It is produced as an alarm pheromone by cheese/mold mites, e.g., *Tyrophagus putrescentiae* (Acari: Acaridae), which are attracted to the cheese because of the production of methyl ketones by the mold.

III. Additional tips for maximizing the usefulness of chemometrics

Several suggestions for maximizing your chances for success with chemometrics are cited:

- Use good peak deconvolution software when using GC-MS peak area/concentration data as independent variables. The advantage that good peak deconvolution software can provide when combined with chemometrics cannot be overstated. One instrument that, in my experience, works particularly well in this regard is the Leco Pegasus III GC-TOFMS system from Leco Corp. (St. Joseph, MO). Not only does it seamlessly and accurately perform GC peak deconvolution, but it also allows rapid and easy manipulation of analytical result tables into spreadsheets suitable for exporting into chemometrics software. It is not uncommon to find 25% more analytes that are coeluting with other analytes using this system. When peak deconvolution software is not employed, there is a good possibility of missing important analytes, and it is also highly likely that inaccurate determination of peak areas for many analytes used in chemometric profiling will be made. The more accurate the chemical data (both qualitatively and quantitatively), the greater the likelihood of resolving problems. The advantage offered by sophisticated GC-TOFMS instruments can easily be worth the relatively high cost of the instrumentation when the instrumentation is applied to important chemometric applications.

  The advantages that TOF instruments offer over quadruple mass spectrometers for flavor and fragrance analysis have been pointed out previously [22]. Quadrupole instruments are scanning instruments that produce mass spectra by sequentially measuring the intensity of individual m/z's over the range selected. By this method, a full mass spectrum is obtained over a period of time, which becomes the interval defining the resolution along the time axis of the resulting chromatograms. For quadruple instruments, the maximum rate is from 5 to 10 spectra/sec with intervals of 0.2 to 0.1 sec.

Scanning over a limited *m/z* range can improve the resolution on the time axis in cases where mass spectral information can be sacrificed.

Array detectors like TOF mass spectrometers do not scan but rather measure all of the ions across the *m/z* range simultaneously. Because all the ions present in the source are simultaneously extracted and subsequently measured, spectra are produced that are entirely free of skewing. Furthermore, because no scanning action is involved, the time required to produce a mass spectrum is greatly reduced. The maximum rate with adequate sensitivity is from 50 to 200 spectra/sec, providing resolutions along the time axis of 0.02 to 0.005 sec. TOF acquisition rates can be increased to 500 complete spectra/sec with a resulting resolution of 0.002 sec on the time axis. The combination of lack of skewing and the rapid mass spectra generation rates make TOF a superior mass analyzer for accurate and sensitive peak deconvolution applications compared to scanning mass analyzers.

• Use the appropriate analytical sample-extraction procedure. Chemometrics is not magic; it is only as good as the quality of analytical data that are used. If the analytes that model the problem you are studying are not extracted by the analytical techniques employed, chemometrics will not help you solve any problems. If fatty acids or carboxylic acids are important independent variables for your particular type of samples, you will want to think of using SPME instead of static headspace or purge-and-trap sample prep techniques. In the case of dairy product analysis, these compounds are extremely good indicators of flavor (desirable and undesirable); SPME can detect these critical compounds, while static headspace and purge-and-trap are poor techniques to analyze these flavor-impact chemicals. If you use an inappropriate SPME fiber for your particular analytes and sample matrix, chemometric results will not be useful.

There is no such thing as a perfect sample preparation technique. All techniques have biases, prejudices, and weaknesses. Because of heavy workload demands, most flavor chemists do not have sufficient time to conduct experiments to optimize analytical sample preparation methods for the problem under study. The specific analytes and matrix components in the samples being studied are important factors in determining the most appropriate sample preparation techniques for a particular problem being studied. Do not assume the sample preparation method that provides the most GC peaks is the best one to use. You are not after quantity as much as quality and, in this case, quality analytes are those that are responsible for the flavor attributes you are trying to understand.

- Do not misuse chemometrics. It is common for chemists unfamiliar with the use of statistical tools to misuse them. We need to be particularly careful not to make assumptions based on chemometrics until we carefully test the validity of our assumptions.

  If your chemometric studies show chemicals A, B, C, and D are critical predictors of flavor scores of a food product, validate this with model system studies. This technique, which is also commonly applied to olfactometry studies, involves preparing a solvent blank, spiking numerous samples of the blank with various levels of the key character-impact chemicals you have identified with chemometrics (or olfactomery experiments), and then subjecting the samples to sensory paneling for evaluation. This is one approach that can be used, for example, when you want to verify that the chemicals you have identified by examining loadings results are, indeed, the most important analytes for causing a particular off-flavor in a product.

  This approach is useful for finding chemicals that appear to be key analytes for modeling specific sample classes but actually only indirectly correlate with the problem. For example, you may have identified pentanal and hexanal as the key contributors to modeling, but the real contributor may be 2-pentylfuran, which may not have been detected by your analytical method. All are oxidation products of linoleic acid, so it is not unusual that they would be present simultaneously in samples. The problem is that 2-pentylfuran (which has a much lower odor-detection threshold than pentanal and hexanal) is the direct indicator of the off-flavor problem you are studying — not pentanal or hexanal. The specific breakdown pathways of the fat hydroperoxides determine which lipid oxidation products are formed. In some cases, pentanal and hexanal may be formed but little 2-pentylfuran. In other cases, 2-pentylfuran may form in disproportionately large concentrations compared to pentanal and hexanal.

- Do not readily accept statistical results that contradict your knowledge about your analytical system. Do not be too quick to replace common sense and experience with statistical results.

- Develop good chemometric habits. In their excellent book *Chemometrics: A Practical Guide*, Kenneth Beebe et al. list six habits of an effective chemometrician [23]. These habits provide a strategy for systematically evaluating data regardless of the method being used:

  Habit 1: Examine the data. Once the data set is collected, visually examine data plots and tables. The purpose of this step is to use the human eye to look for obvious errors or features in the data. Because errors can occur in both the measurement variables or characteristic values, it is important to examine both of these sets of numbers in this step.

Habit 2: Preprocess as needed. There can be random or systematic sources of variation that mask the variation of interest. This unwanted variation may reduce the effectiveness of the model. An understanding of the chemistry underlying these unwanted sources of variation helps with appropriate selection of preprocessing techniques. Use preprocessing with caution. Remember that preprocessing changes the data set and, if inappropriately applied, can remove important variation from the data.

Habit 3: Estimate the model. The next step is to generate the chemometric model and associated diagnostics.

Habit 4: Examine the results/validate the model. All of the chemometric methods generate numerical and graphical results. You should carefully examine the computer output with the goal of validating the model (i.e., determining if the model is reliable). Diagnostic tools for each of the methods are used to assess the confidence that can be placed in the results. If the model is unacceptable, refinement is often possible by adjusting the model parameters.

Habit 5: Use the model for prediction. For methods that generate a predictive model, Habit 5 is the ideal model application for unknown samples. The output from this habit is the predicted properties or classes of the unknown samples. With PLS, for example, calibration models are constructed to predict properties of future samples.

Habit 6: Validate the prediction. Chemometrics software rarely fails to make predictions once a model and unknowns are provided in the proper format. It is, therefore, important that all results be validated. The capability to validate prediction results is one of the greatest advantages of using multivariate techniques. Validation increases the chances of making good decisions based on the outputs from the models by indicating the confidence that should be placed on the predicted values.

Acknowledgments

The chemometrics software used for most of the examples presented in this chapter is Pirouette (Infometrix, Bothell, WA). Also, Scott Ramos, Paul Bailey, Brian Rohrback, and others at Infometrix, Inc., have been an excellent source of chemometrics advice and support for me for many years.

References

1. Infometrix, Bothell, WA, www.infometrix.com.
2. R.T. Marsili, Combining mass spectrometry and multivariate analysis to make a reliable and versatile electronic nose, *Flavor, Fragrance and Odor Analysis*, R. Marsili, Ed., Marcel Dekker, New York, 2002, p. 349.
3. V.R. Kinton, R.J. Collins, B. Kolahgar, and K.L. Goodner, Fast Analysis of Beverages using a Mass Spectral Based Chemical Sensor, Gerstel Application Note 4/2003.

4. V.R. Kinton J.A. Whitecavage, A.C. Heiden, and C. Gil, Use of a Mass Spectral Based Chemical Sensor to Discriminate Food and Beverage Samples: Olive Oils and Wine as Examples, Gerstel Application Note 1/2004, Baltimore, MD.

5. V.R. Kinton, E.A. Pfannkoch, M.A. Mabud, and S.M. Dugar, Wine Discrimination Using a Mass Spectral Based Chemical Sensor, Gerstel Application Note 2/2003, Baltimore, MD.

6. B. Kolahgar and A.C. Heiden, Discrimination of Different Beer Sorts and Monitoring of the Effect of Aging by Determination of Flavor Constituents Using SPME and a Chemical Sensor, Gerstel Application Note 11/2002, Baltimore, MD.

7. R.T. Marsili, SPME-MS-MVA as a rapid technique for assessing oxidation off-flavors in foods, *Headspace Analysis of Foods and Flavors: Theory and Practice/ Advances in Experimental Medicine and Biology*, Vol. 488, R.L. Rouseff and K.R. Cadwallader, Eds., Kluwer Academic/Plenum Publishers, New York, 2001, p. 56.

8. R.T. Marsili, Flavors and off-flavors in dairy foods, in *Encyclopedia of Dairy Science*, H. Roginski, J.W. Fuquay, P.F. Fox, Eds., Academic Press, London, 2003, pp. 1069–1081.

9. R.T. Marsili, Shelf life prediction of processed milk by solid-phase microextraction, mass spectrometry and multivariate analysis, *J. Agric. Food Chem.* 48: 3470–3475, 2000.

10. R.T. Marsili and N. Miller, Determination of the cause of off-flavors in milk by dynamic headspace GC-MS and multivariate data analysis, *Food Flavors: Formation, Analysis and Packaging Influences*, E.T. Contis, C.-T. Ho, C.J. Mussinan, T.H. Parliment, F. Shahidi, and A.M. Spanier, Eds., Elsevier Science B.V., The Netherlands, 1998, p. 159.

11. R.T. Marsili and N. Miller, Off-flavors in milk by dynamic headspace GC-MS and multivariate data analysis, *Food Flavors and Chemistry: Advances of the New Millennium*, A.M. Spanier, F. Shahidi, T.H. Parliament, C. Mussinan, E.T. Contis, and C.-T. Ho, Eds., Elsevier Science B.V., The Netherlands, 2001, p. 118.

12. D.A. Guillen, M. Palma, R. Natera, R. Romero, and C.G. Barroso, Determination of the age of Sherry wines by regression techniques using routine parameters and phenolic and volatile compounds, *J. Agric. Food Chem.* 53: 2412–2417, 2005.

13. V.A. Watts, C.E. Butzke, and R.B. Boulton, Study of aged cognac using solid-phase microextraction and partial least-squaress regression, *J. Agric. Food Chem.* 51: 7738–7742, 2003.

14. M.S. Perez-Coello, P.J. Martin-Alvarez, and M.D. Cabezudo, Prediction of the storage time in bottles of Spanish white wines using multivariate statistical analysis, *Z. Lebensm.-Unters. Forsch. A.* 208: 408–412, 1999.

15. J.F. Clapperton and J.R. Piggott, Differentiation of ale and lager flavors by principal components analysis of flavor characterization data. *J. Inst. Brew.* 85: 271–274, 1979.

16. T. Jacobsen and R.W., Gunderson, Cluster analysis of beer flavor components. II. A case study of yeast strain and brewery dependency, *J. Am. Soc. Brew. Chem.* 41: 78–80, 1983.

17. T. Jacobsen, R. Volden, S. Engan, and O. Aubert, A chemometric study of some beer flavor components, *J. Inst. Brew.* 85: 265–270, 1979.

18. L.E. Stenroos and K.J. Siebert, Application of pattern recognition techniques to the essential oil of hops, *J. Am. Soc. Brew. Chem.* 42: 54–61, 1984.

19. K.J. Siebert and L.E. Stenroos, The use of multivariate analysis of beer aroma volatile compound patterns to discern brand-to-brand and plant-to-plant differences. *J. Am. Soc. Brew. Chem.* 7: 93–101, 1988.

20. M. Moll, V. That, R. Flayeux, and P. Muller, Prediction of the organoleptic quality of beer, in *The Quality of Foods and Beverages,* Charalambous, G. and Inglett, G., Eds., Academic Press: New York, 1981, pp. 147–166.

21. J.-H. Lee, R. Diono, G.-Y. Kim, and D.B. Min, Optimization of solid phase microextraction analysis for the headspace volatile compounds of Parmesan cheese, *J. Agric. Food Chem.* 51: 1136–1140, 2003.

22. J.F. Holland and B.D. Gardner, The advantages of GC-TOFMS for flavor and fragrance analysis, *Flavor, Fragrance and Odor Analysis,* R. Marsili, Ed., Marcel Dekker, New York, 2002, p. 107.

23. K.R. Beebe, R.J. Pell, and M.B. Seaholtz, *Chemometrics: A Practical Guide*, John Wiley & Sons, New York, 1998, p. 3.

chapter eight

Sensometrics: the application of multivariate analysis to sensory data

S. Karow, Y. Fu, and T. Laban

Contents

I. Introduction

Multivariate analysis is a tool used by the sensory professional to simplify and aid in the interpretation of descriptive, consumer, and analytical data. Multivariate techniques are mainly used in data reduction (Principal Component Analysis [PCA], Factor Analysis [FA], Correspondence Analysis, etc.), for classification (FA, hierarchical cluster analysis, discriminant analysis, etc.), and data relationships (multiple regression, principal component regression, partial least square analysis (PLS)) (Piggott, 1986). In the industry, it can be used in product development to guide R&D reformulations, aid in understanding process variation (e.g., plant-to-plant, line-to-line, or within line or scale-up), monitor and identify panelist performance issues, shelf life studies, etc. Multivariate analysis can also be used to understand and identify key attributes driving consumers' liking, segmentation (e.g., age, gender, geographic region), and reducing the number of attributes on a ballot. The opportunities are endless. In practice, sensory professionals have ample opportunities every day to use multivariate techniques to understand the sensorial world around them.

The objective of multivariate reduction techniques is to create concise variables that contain most of the information present in the original multi-dependent variables. As a result, multivariate techniques allow complex data sets to be collected from several sources and deliver the results in easy-to-interpret tables and charts. PCA is performed in order to simplify the description of a set of interrelated variables. It can be summarized as a method that transforms the original variables into new uncorrelated variables, which are called principal components (PC). PCA finds linear combinations that will maximize the total variance. Regression coefficients are estimated such that within a PC total, variation is maximized and each is independent.

The following chapter provides several examples of how sensory scientists can use multivariate analysis to understand sensory differences between samples. In all of the examples, the Sensory Spectrum method of descriptive analysis was used to evaluate the samples. This method uses highly trained sensory judges who evaluate products using a 15-point intensity scale (0 = none, 15 = extreme). Reference materials including basic taste solutions were available at each session to calibrate the panelists.

The first study was conducted to help QC identify an analytical quality measurement that could accurately predict the sensory properties of cinnamon-flavored cereal. The hypothesis was that product color would be a good indicator of product quality. The data was analyzed using XLSTAT software. The second project was conducted to determine whether two production lines produced the "same" cheese sauce product. In this case, both descriptive and analytical data was collected for over 12 months from both lines and analyzed using SPSS multivariate software. In the third study, descriptive analysis was used to determine key flavor and texture attributes of chocolates (bittersweet, semi-sweet, low-carb, and milk chocolate). The data was analyzed using Minitab multivariate analysis software. Again, these are examples of how multivariate data can be used to analyze and interpret large data sets.

II.   The use of multivariate analysis in examining sensory data: application examples

A.   Cinnamon-flavored cereal

The following study was conducted to better understand the impact of moisture levels on the sensory and analytical parameters of cinnamon-flavored cereal. The current moisture specifications for cinnamon-flavored cereal ranges from 2 to 4% moisture. These results were analyzed using XLSTAT.

1.   Materials and methods

One hundred and thirty samples of granola cereal were collected from a single production line over a 19-h shift. Ten samples were collected every 30 min and submitted for both moisture analysis and descriptive analysis to characterize and quantify any appearance, flavor, and texture differences.

Samples were systematically evaluated using descriptive analysis techniques for appearance, flavor, and texture. A total of 12 descriptors were used for the analysis. The descriptors are listed in Table 8.1. Samples were initially evaluated for visual characteristics to determine the impact of moisture on appearance. Over the entire range of samples, only slight differences were noted in color and appearance making it impossible to reliably evaluate these attributes. Samples were then evaluated to determine the impact of moisture on flavor and texture.

At the lower moisture range (2.00 to 2.24%), the samples had distinct burnt notes, reduced hardness, and molar packing. In the mid range (2.25 to 3.15%), the samples were toasted and sweet with noticeable caramelized notes. At the

Table 8.1 Cereal Attributes

Overall flavor
Cinnamon flavor
Toasted
Balanced flavor
Burnt
Raw oats
Sweet
Bitter
Fracturability
Molar packing
Rate of breakdown
Denseness

high range (3.46 to 4.30%), the samples had high levels of raw-oat flavor, low caramelized flavor, hard texture, and increased molar packing.

2. Results

Analytical data indicated samples collected over the 19-h shift fell within specifications with percent moistures ranging from 2.00 to 4.00%. Descriptive analysis showed that perceivable differences in flavor and texture could be detected at 0.3% moisture of the cinnamon-flavored cereal.

PCA was conducted to summarize the overall differences of the cinnamon-flavored cereal from the 19-h production run. Two principal components (PC) explained 93% of the variation. PC1, which mainly separated samples based on moisture, explained 69% of the variation. PC2, which separated samples based on the overall flavor, accounted for 24% of the variation. High-moisture cinnamon-flavored cereals were positively correlated, and low-moisture cinnamon-flavored cereals were negatively correlated with the PC1 axis. The attributes for both PC1 and PC2 are shown in Table 8.2. Figure 8.1 shows the correlation circle and relationships between attributes of the cinnamon-flavored cereal. Positive PC1 was highly correlated with cinnamon, sweet, and denseness and negative PC1 was largely correlated with toasted, fracturability, and breakdown. Figure 8.2 shows the PCA plot of the cinnamon-flavored cereals and attributes. The plot illustrates high-moisture cinnamon-flavored cereals were positively correlated to PC1, and low-moisture cinnamon-flavored cereals were negatively correlated to PC2. Positive PC2 was highly correlated with overall flavor and negative PC2 was highly correlated with molar packing.

3. Conclusions

Based on the findings, it can be concluded that percent moisture is the key source of differences between the cinnamon-flavored cereals. Five distinct moisture ranges were identified, with each producing perceivably different products.

Table 8.2 Factor Loadings (XLSTAT)

	F1	F2	F3	F4	F5
Overall flavor	−0.057	0.928	0.302	0.210	0.022
Cinnamon	0.989	0.085	0.060	0.056	0.088
Bitter	−0.701	0.685	−0.087	0.176	−0.025
Toasted	−0.977	−0.139	0.039	0.152	−0.038
Sweet	0.986	−0.059	−0.052	0.144	0.024
Flavor balance	−0.740	0.596	0.176	0.258	0.003
Burnt	−0.834	−0.540	−0.102	0.039	−0.044
Raw oat flavor	0.916	0.278	0.075	0.256	−0.116
Fracturability	−0.969	−0.001	−0.070	0.176	0.159
Molar packing	−0.032	−0.892	0.450	0.017	0.013
Breakdown	−0.977	−0.139	0.039	0.152	−0.038
Denseness	0.986	−0.059	−0.052	0.144	0.024

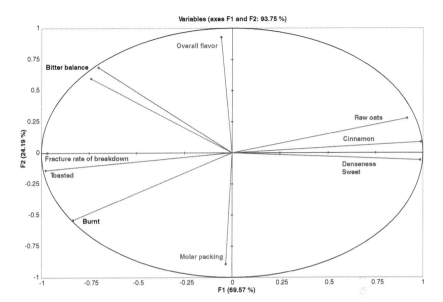

Figure 8.1 Correlation circle (XLSTAT).

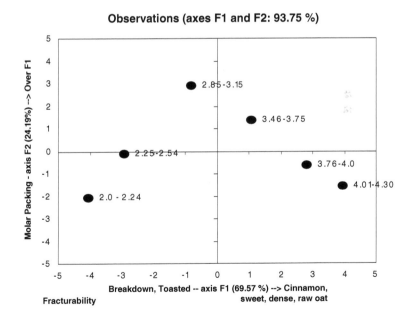

Figure 8.2 PCA plot (XLSTAT).

Because the products could not be differentiated by visual means, another method had to be developed to identify differences between the samples. In this case, multivariate analysis identified percent moisture as the major contributor to sample differences. Analytical procedures were instituted to track percent moisture on a regular basis.

4.   Looking to the future

The next steps include conducting an experimental design to identify the key ingredients and process conditions (e.g., bake time, burner temperature, and air flow), which have the greatest impact on the moisture of the finished product. Finally, an optimization study will be conducted on these key contributors to identify the optimal levels and settings. These steps will reduce product variation and ensure more consistent products over time. As part of this follow-up study, consumers can be asked to evaluate the products to help set product specifications.

B.   Cheese sauce

Statistics tools such as ANOVA, MANOVA, correlation, regression, and PCA are commonly used in academic fields to build models for research. This study is an example of how statistical tools can be implemented in a day-to-day quality assurance program in the food industry to monitor and make timely adjustments to key parameters.

One of the challenges many manufacturers face is maintaining consistent quality in a product. The quality of cheese sauce is affected by various factors, such as raw-material quality, season and year variation, operation parameters, and storage conditions. In this study, cheese sauce was produced using either a batch or continuous process line. Management's objective was to use the lines interchangeably, based on product demand, and also maintain set specifications and similar product profiles. The following study was conducted using historical data to test the theory that the two production lines manufactured similar products.

As part of the validation process, both analytical and sensory data were collected from two different production lines (A and B) over approximately 1 year. Initial univariate statistics were performed to understand process-control variations. Results indicated that all samples appeared to be similar, based on "within specification" for all analytical, sensory, and six sigma. Variables were then analyzed simultaneously using univariate and multivariate analysis. The advantage of multivariate analysis is the ability to detect broader patterns of interrelationships among products and among sensory and analytical characteristics compared to individual univariate analyses.

1.   Materials and methods

A total of 173 cheese-sauce samples were selected. They were manufactured on two different production lines (A and B) within a 12-month period. A gold standard was carefully selected based on overall quality (sensory and analytical)

Table 8.3 Sensory and Analytical Attributes

Sensory:
  Salty
  Sour
  Cheddar
  Dairy
  Sweet
  Bitter
  Soapy
  Overall
  Viscosity
  Acidic milk

Analytical:
  Fat
  Salt
  Color
  L COLOR
  a COLOR
  b COLOR

and was used in this study as the control sample. All gold standard samples were stored under frozen storage conditions in amber jars.

A panel consisting of eight trained and well-calibrated judges evaluated 173 cheese-sauce samples using the spectrum-descriptive technique. Within each set, blind control samples were used to monitor the panelist's performance. Instrumental results were obtained from the analytical lab at the plant. Table 8.3 lists the sensory and analytical attributes which were collected in this study.

2. Results

ANOVA (Table 8.4) showed that sour, TA, FAT, and bCOLOR were the parameters differentiating the cheese-sauce samples made from two process lines. Correlation (Table 8.5 and Table 8.6) suggested that salty, sour, and overall

Table 8.4 ANOVA Results (SPSS)

Attribute	Sour 0.010*	TA 0.045	FAT 0.050	b COLOR 0.012
A	3.31[b]**	19.38[b]	31.81[b]	35.49[a]
B	3.15[ab]	18.37[a]	31.61[b]	35.68[b]
Gold Standard	3.07[a]	18.90[a]	31.2[a]	35.4[a]

Note: P-values: all attributes listed are significantly different (< 0.05) from products made on production line A and B.

** Within the same attribute, the means with the same letter (a, b) are not significantly different from each other.

Table 8.5 Correlation Results (SPSS)

Correlation	Salty	Sour	Dairy	Cheddar	Sweet	Bitter	Soapy	SALT	TA	L COLOR
Salty										
Sour	**0.202**									
Dairy	-0.172	-0.187								
Cheddar			0.152							
Sweet		-0.169	**0.227**							
Overall	0.225	0.158		0.153						
Bitter										0.166
Soapy				-0.388						
Viscosity										
Acidic Milk	-0.238	0.17								
FAT										
SALT										
TA		0.353			-0.212		0.16			
L COLOR						0.166				
a COLOR					0.158	-0.231		-0.157	-0.25	-0.727
b COLOR		-0.156								0.409

Correlation is significant at the 0.05 level (2-tailed).

Correlation is significant at the 0.01 level (2-tailed).

Table 8.6 PCA Factors (SPSS)

Component	Factor 1	Factor 2	Factor 3
Variance%	21.1	20.7	15.1
Salty	0.530	−0.402	−0.161
Sour	0.291	−0.526	0.176
Dairy	0.061	0.754	0.032
Cheddar	0.243	0.215	−0.744
Sweet	0.090	0.636	−0.121
Overall	0.843	0.196	0.128
Soapy	0.106	0.012	0.868

flavor; sweet and dairy; soapy and cheddar (reversely); TA and sour; aCOLOR and bitter (reversely); bCOLOR and LCOLOR are significantly correlated with each other and therefore possibly can be used as predictors for each other. Figure 8.3 shows the PCA mapping where more products made from Line A had high sour and salty ratings, and those made from Line B had high soapy flavor. The majority of the products made from both lines had medium to high cheddar, sweet, and dairy notes. Linear regression (Table 8.7) was conducted to reveal the contribution of all the parameters to the deviation scores (define or explain deviation scores). Cheddar, Viscosity, Salt, and TA were suggested to be significant contributors and should therefore be considered as the key quality-control aspects in future production processes.

3.   Conclusions

In this study, multivariate methods (correlation, PCA, regression) were applied to sensory and analytical data derived from cheese sauces produced by different processes. Using multivariate analysis, related variables were

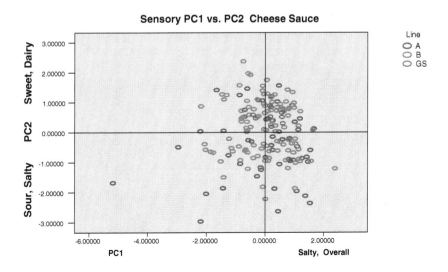

Figure 8.3 Sensory PC1 vs. PC2 cheese sauce (SPSS).

Table 8.7 Linear Regression (SPSS)

	Unstandardized coefficients	Standard error	Standardized coefficients beta	t	Sig.
Constant	10.54	13.74		0.77	0.44
Cheddar	0.38	0.20	0.15	1.92	0.06
Viscosity	−0.22	0.09	−0.16	−2.31	0.02
SALT	−1.08	0.38	−0.21	−2.87	0.00
TA	−0.21	0.09	−0.20	−2.45	0.02

identified, characteristics typical of a particular product from a process line were established, and key variables for predicting quality were determined.

4. Looking to the future

The next step would be a designed experiment further studying the impact of the identified key attributes on cheese-sauce quality. The samples should be evaluated with actual consumers. The ultimate goal is to achieve high-standard cheese-sauce products with consistent quality over time and the ability to predict quality from existing data.

C. Chocolate

For the last example, only the multivariate plots will be presented so that the discussion can focus more on the interpretation of the results, rather than the actual analysis.

In an effort to create a new chocolate with a unique, signature sensory profile, a large-scale competitive evaluation was conducted. A review of current product profiles (qualitative and quantitative) was established based on descriptive sensory evaluation of over 130 chocolate and chocolate-flavored products from around the world. The multivariate approach was used to analyze the data. This approach efficiently summarized the large data set and illustrated broad configurations of variable interrelationships.

1. Materials and methods

Over 130 international chocolate and chocolate-flavored products were initially purchased from local grocery and specialty stores. Subsets of samples with different flavor and textural characteristics were presented to the panelists to show a broad representation of characteristics.

Twelve panelists were trained extensively for 6 weeks on the most common attributes, as well as distinctive characteristics found in the sample sets. Spectrum descriptive analysis method was used. Panelists were taught how to use the 15-point intensity scale (0 = none, 15 = extreme), and corresponding reference materials specific to each attribute were provided. Overall, panelists were exposed to more than 300 samples (at least 80 of which were

Table 8.8 Chocolate Samples

1	Hershey's Milk
2	Russell Stover's — Low Carb
3	Scharffen Berger — 99%
4	Bernard Castelain
5	Hershey's Low Carb
6	Cadbury (Canada)
7	Cadbury (U.S.)
8	Blanxart
9	Lindt Swiss Milk
10	Milka
11	Terra Nostra
12	Droste
13	AdvantEdge
14	El Rey
15	Joseph Schmidt
16	Dagoba
17	Ghirardelli Milk
18	Lake Champlain Belgian Milk
19	Chocolove Milk
20	Dove Milk
21	Sam's Club Milk Chocolate
22	Dagoba 87%
23	Newman's Own
24	Sarotti
25	Godiva
26	Divine
27	Carlos V
28	Russell Stover's Sugar-Free
29	Lindt Extra Dairy
30	Scharffen Berger 62%

nonrepeated), and they identified nearly 70 flavor, visual, and textural attributes to describe the world of chocolates.

Table 8.8 lists the 30 chocolate and chocolate-flavored samples that were evaluated. Table 8.9 shows a list of attributes used for the final evaluation of the chocolate samples.

2.   Results

The descriptive data was summarized using PCA as shown in Figure 8.4. About 50% of the variation between the samples was explained with Factor 1 (29.5%) and Factor 2 (21.6%). Factors 1 to 4 explained 81% of the variation. Only Factors 1 and 2 will be discussed.

Based on the distribution of the attributes and associated products, three major groups were identified: "dark chocolate group" (right) having relatively high flavor intensities in bitter, bitter aftertaste, cocoa, and overall flavor and aroma; "milk chocolate group" (left) characterized by high levels

Table 8.9 Chocolate Attributes

Aroma	Tobacco	Cinnamon	Caramel/Cooked
Overall	Fruity	Chemical	Non-dairy
Appearance	Malt	Grainy/ceral/soy flour	Brown Spice
Color	Floral	Chalk	**Basic Taste**
Shine	Nutty/Woody	Coconut	Sweet
Bloom	Winey/Fermented	Rancid vanilla	Sour
Flavor	Olive	Green olives	Salty
Flavor onset (sec)	Cherry	**Aftertaste**	Bitter
Overall flavor	Citrus	Bitter	**Texture/Feelings**
Chocolate	Coconut	Chocolate	Astringent
Cocoa	Oil	Cocoa	Cooling
Milk/dairy	Burnt	Coffee	Throat burn
Casein	Green	Milky	Tongue burn
Caramel/Cooked	Sugar Alcohol	Sweet	Hardness
Vanilla/Vanillin	Other-Earthy/Dirty	Tobacco	Graininess
Coffee	Ash	Burnt	Onset of melt
Expresso	Butterscotch	Art Sweet	Viscosity
			Adhesive

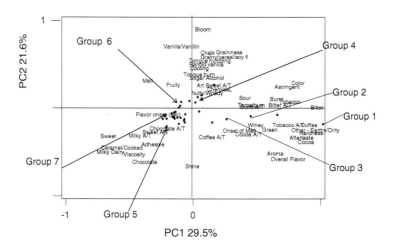

Figure 8.4 Sample positions and attributes: all samples PC1 and PC2 (Minitab).

Table 8.10 Cluster Analysis Dendrogram (Minitab)

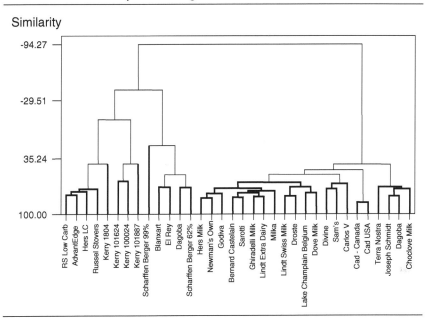

of sweet, milky and milky aftertaste, caramel, cooked, and viscosity; "mouth feel" group (top) associated with high intensities of chalky, graininess, tongue numbing, cooling, tongue burn, as well as bloom and vanilla. Nearly half of the samples were located in the milk-chocolate group.

Using multivariate cluster analysis (Table 8.10), samples were grouped into seven clusters, three of which lie close to the center and could be further analyzed. The group of clusters are further shown and included on the PCA indicated in Figure 8.5.

Group 1 consists of one chocolate: Scharffen Berger 99%. Attributes associated with this sample include: aroma, overall flavor, color, aftertaste, cocoa, coffee, bitter, hardness, astringent, earthy or dirty, burnt, and sour.

Group 2 is similar to Group 1: However, overall intensity is lower than Group 1 in some attributes, comprising of Scharffen Berger 62% and Dagoba 87%. Attributes associated with these samples include: aroma, overall flavor, color, aftertaste, cocoa, bitter, and hardness. These were considered to be among the most extreme chocolates evaluated (darker, more bitter chocolates), which is indicated by their location on the far right of Figure 8.4.

Group 3 consists of El Rey and Blanxart products. Attributes are similar to the previous two groups but are different in intensity and display a move towards characteristics on the left.

Four products form Group 4: AdvantEdge, Russell Stover's Low-Carb, Hershey's Low-Carb, and Russell Stover's Sugar-Free. These samples are located close to the center of the plot and are not explained well. Analysis suggests

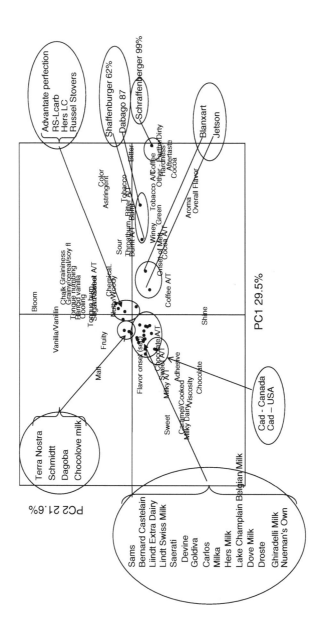

Figure 8.5 Cluster groupings: all samples PC1 and PC2 (Minitab).

that they have: artificial sweet, cooling, sugar alcohol, some shine, slight coconut, nutty or woody, and are low in adhesive and viscosity.

Group 5 comprises two Cadbury samples having characteristics of high aroma, flavor, shine, milky or dairy, caramel, aftertaste, sweetness, graininess, viscosity and adhesive, and are low in bitterness.

Group 6 consists of Terra Nostra, Joseph Schmidt, Dagoba, and Chocolove milk products. These samples are located close to the center of the plot. Analyses suggest that they have: shine, chocolate, sweetness, graininess, a slightly lower viscosity, and are low in aftertaste and bitterness.

Group 7 is by far the largest group with 15 samples, and further analyses subdivide this. Samples tended to be sweeter, exhibited more chocolate flavor, were lighter in color, and low in bitterness and graininess.

Figure 8.6 shows further separation of Group 7. The plot illustrates clustering of similar samples among the group.

The multivariate cluster-analysis results incorporating all products isolated groups that appeared to be similar. The individual clusters were then analyzed independently to further refine interrelationships among the clusters. Additional distinguishing relationships were uncovered, making a convincing case for a multistep, multivariate analysis. That is, when based upon the multiple attributes of the group, distinctions within the group become clearer.

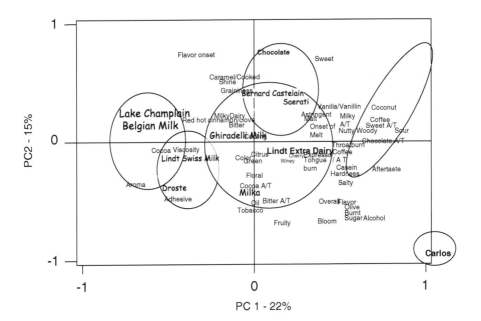

Figure 8.6 Fifteen-sample milk chocolate cluster PC1 and PC2 (Minitab).

Table 8.11 Cluster Analysis Dendrogram 15 Sample Cluster (Minitab)

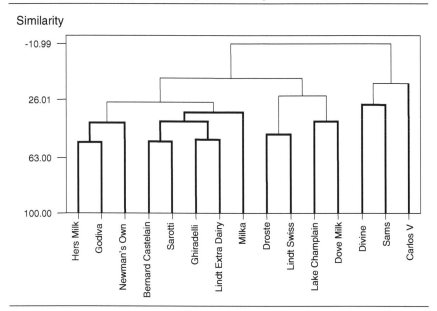

3. Conclusions

Using various multivariate analyses techniques, analysts uncovered a large representation of chocolate profiles. The relationship of the products in "space" was illustrated, the characteristics that drive their own particular profile or differentiate them from others was determined, and the potential to develop a new, unique chocolate with its own "sensory identity" was explored.

The PCA plot (Figure 8.5) is a virtual map of the different products; i.e., it tells us where products lie in relation to each other. Whereas some samples have attributes that place them at the extremes of the map, such as tobacco or bloom, others are clustered closer to the origin. The distance between two such products on the map would be an indication of the degree of their similitude or lack thereof.

The cluster analysis tells a similar story. Table 8.11 gives approximate percentage values to the similarity of the products.

4. Looking to the future

The next steps in our research will involve combining large-scale consumer (n = 200+) testing with descriptive and analytical testing. Using the results from the current study, one or two products from each cluster grouping will be selected for further testing. The analytical and descriptive data will objectively quantify the differences between the samples, and the consumer testing will quantify consumers' acceptance and preference for each sample. The consumer data will be analyzed to identify segmentation (age, gender, etc.)

and geographic preferences. By cross-referencing these results with those from our own evaluations, we can begin to understand which attributes significantly affect consumer-purchase behavior. In the future, we would also like to repeat these studies in Europe, South America, and Central America to establish global preferences and trends for types of chocolate and chocolate products. By including variables such as geographic region, gender, and age, the benefits of using multivariate analysis become even more apparent.

References

Lawless, H.T. and Heymann, H. 1999, *Sensory Evaluation of Foods: Principles and Practices*, Chapman and Hall Food Science, Gaithersburg, MD.

Meilgaard, M. et al., *Sensory Evaluation Techniques*, 3rd ed., CRC Press, Boca Raton, FL, 1999.

Minitab, Version 14.0. Minitab Inc., State College, PA.

Piggott, J.R., *Statistical Procedures in Food Research*, Elsevier Applied Science, London, 1986.

Richarme, M., Eleven Multivariate Techniques: Key Tools in Your Marketing Research Survival Kit, Decision Analysis, 2001.

Rittman, A.C.R.C. 2005. From Cheese to Cheese Sauce, in *Food Product Design*, Culinary, March 2005.

SPSS for Windows, Advanced analysis and graphic packages, Version 14, Information Technology Services, Austin, TX.

XLSTAT software, Addinsoft, Inc., New York.

chapter nine

Character-impact flavor compounds

Robert J. McGorrin

Contents

I. Introduction

Aroma substances that comprise food flavors occur in nature as complex mixtures of volatile compounds. However, a vast majority of volatile chemicals that have been isolated from natural flavor extracts do not elicit aroma contributions that are reminiscent of the flavor substance. For instance, *n*-hexanal is a component of natural apple flavor [1], however when smelled in isolation, its odor is reminiscent of "green, painty, rancid oil." Similarly, ethyl butyrate provides a nondescript "fruity" aroma to blackberries, raspberries, and pears, but it does not distinctly describe the flavor quality of any of these individual fruits. It has long been the goal of flavor chemists to elucidate the identity of pure aroma chemicals that possess the unique flavor character of the natural fruit, vegetable, meat, cheese, or spice that they were derived from. Frequently, these unique flavor substances are referred to as "character-impact compounds" [2].

The character-impact compound for a particular flavor or aroma is a unique chemical substance that provides its principal sensory identity. Often, character impact is elicited by a synergistic blend of several aroma chemicals. When tasted or smelled, the character-impact chemical, or a group of chemicals, contributes a recognizable sensory impression even at the low concentration levels typically found in natural flavors (for example, vanillin in vanilla extract and diacetyl in butter) [3,4]. In some instances, flavor concentration and food context are very important. For example, at high concentrations, 4-mercapto-4-methyl-2-pentanone ("cat ketone") has an off-odor associated with cat urine, but when present at reduced levels in the context of a Sauvignon blanc wine, it conveys the typical flavor impression of the Sauvignon grape [5].

For many foods, character-impact flavor compounds are either unknown or have not been reported to date. Examples of these include cheddar cheese, milk chocolate, and sweet potatoes. For these foods, the characterizing aroma appears comprise a relatively complex mixture of flavor compounds, rather than one or two aroma chemicals.

The intent of this chapter is to summarize and update a previous review [6] regarding what is generally known about the chemical identities of characterizing aroma chemicals in fruits, vegetables, nuts, herbs and spices, savory, and dairy flavors. A brief compendium of characterizing off-flavors and taints that have been reported in foods is also discussed.

II.   Character-impact flavors in foods

More than 6000 compounds have been identified in the volatile fraction of foods [7]. The total concentration of these naturally occurring components varies from a few parts-per-million (ppm) to approximately 100 ppm, with the concentration of individual compounds ranging from parts per billion to parts per trillion. A majority of these volatile compounds do not provide significant impact to flavor. For example, more than 800 compounds have been identified in the flavor of coffee, but in general only a small proportion of these substances have a significant contribution to its sensory flavor profile [8].

The ultimate goal of flavor research is to identify and classify unique aroma chemicals which contribute to the characteristic odor and flavor of foods. Having this knowledge enables the flavor industry to better duplicate flavors through nature identical or biosynthetic pathways, and can facilitate better quality control of raw materials by screening of the appropriate analytical target compounds.

In recent studies, potent aroma compounds have been identified using various gas chromatography-olfactometry (GC-O) techniques, such as Charm Analysis™ and aroma extract dilution analysis (AEDA) [9,10]. The flavor compounds that are identified by these methods are significant contributors to the sensory profile. In some cases, these sensory-directed analytical techniques have enabled the discovery of new character impact compounds. However, in other instances, key aroma chemicals have been identified which, while potent and significant to flavor, do not impart character impact. For example, in dairy products, chocolate, and kiwifruit, these flavor types appear to be produced by a complex blend of noncharacterizing key aroma compounds.

When character-impact compounds are known, flavor chemists are able to use these materials as basic "keys" to formulate enhanced versions of existing flavors. As analytical techniques improve in sensitivity, flavor researchers continue their quest to discover new character-impact flavors that will enable them to develop the next generation of improved flavor systems.

A. Herb, spice, and seasoning flavors

The original identifications of character aroma compounds were from isolates of spice oils and herbs. Many of these early discoveries paralleled developments in synthetic organic chemistry [11]. The first identifications and syntheses of character flavor molecules include benzaldehyde (cherry), vanillin (vanilla), methyl salicylate (wintergreen), and cinnamaldehyde (cinnamon). A listing of character-impact compounds found in herb and spice flavors is presented in Table 9.1.

A major contributor to the flavor of basil is methyl chavicol (estragole), which provides tealike, green, hay, and minty notes [12]. In Italian-spice dishes, basil is complemented by oregano, for which carvacrol is a character-impact aroma. Thyme is the dried leaves of *Thymus vulgaris,* a perennial of the mint family, for which thymol provides a warm, pungent, sweetly herbal note, with contributions from carvacrol [13]. Interestingly, with fennel (*Foeniculum vulgare*) seed and oil, the essential flavor character is not a single primary compound, but is provided by a combination of *trans*-anethole (anise, licorice), estragole (basil, anise), and fenchone (mint, camphor) [13]. In a classic example of the effect of chiral isomers on flavor character, (S)-(+)-carvone imparts caraway, whereas (R)-(–)-carvone provides spearmint flavor [14].

Toasted and ground fenugreek seed is an essential ingredient of curry powders. Sotolon (3-hydroxy-4,5-dimethyl-2(5*H*)furanone) has been established as a character-impact flavor component in fenugreek on the basis of its "seasoning-like" flavor note [15]. Its aroma characteristic changes from caramel-like at low concentration levels to currylike at high concentrations. The sensory impact of sotolon is attributed to its extremely low detection threshold (0.3 mg/l water), and the fact that its concentration in fenugreek seeds is typically 3000 times higher than its threshold.

Turmeric is also primarily used as a spice component in curry dishes, and as a coloring agent in dried and frozen foods. The character-impact compound for turmeric is reported as *ar*-turmerone [16]. Saffron, the dried red stigmas of *Crocus sativus* L. flowers, is utilized to impart both color and

Table 9.1 Character-Impact Compounds in Herbs, Spices, and Seasonings

Character-Impact Compounds	CAS Registry No.	Occurrence	Reference
trans-p-Anethole	[4180-23-8]	Anise	4
Methyl chavicol (estragole)	[140-67-0]	Basil	12
(S)-(+)-Carvone	[2244-16-8]	Caraway	14
trans-Cinnamaldehyde	[104-55-2]	Cinnamon	4
Eugenol	[97-53-0]	Clove	4
Eugenyl acetate	[93-28-7]	Clove	12
d-Linalool	[78-70-6]	Coriander	4
trans-2-Dodecenal	[20407-84-5]	Coriander	18
Cuminaldehyde	[122-03-2]	Cumin	12
p-1,3-Menthadien-7-al	[1197-15-5]	Cumin	12
(S)-α-Phellandrene	[99-83-2]	Fresh dill	12
3,9-Epoxy-*p*-menth-1-ene	[74410-10-9]	Fresh dill	20
1,8-Cineole (eucalyptol)	[470-82-6]	Eucalyptus	14
Sotolon	[28664-35-9]	Fenugreek	15
Benzenemethanethiol	[100-53-8]	Garden cress	25
Diallyl disulfide	[2179-57-9]	Garlic	21
Diallylthiosulfinate (allicin)	[539-86-6]	Garlic	21
1-Penten-3-one	[1629-58-9]	Horseradish	4
4-Pentenyl isothiocyante	[18060-79-2]	Horseradish	21
Allyl isothiocyanate	[57-06-7]	Mustard	4
Propyl propanethiosulfonate	[1113-13-9]	Onion, raw	21
3-Mercapto-2-methylpentan-1-ol	[227456-30-6]	Onion, raw	24
Allyl propyl disulfide	[2179-59-1]	Onion, cooked	21
2-(Propyldithio)-3,4-Me$_2$thiophene	[126876-33-3]	Onion, fried	21
Carvacrol	[499-75-2]	Oregano	12
α-*t*-, *c*-Bergamotene	[6895-56-3]	Pepper, black	19
l-Menthol	[89-78-1]	Peppermint	4
t-4-(MeS)-3-butenyl isothiocyanate	[13028-50-7]	Radish	21
Verbenone	[80-57-9]	Rosemary	14
(R)-(−)-Carvone	[6485-40-10]	Spearmint	14
Safranal	[116-26-7]	Saffron	17
ar-Turmerone	[532-65-0]	Turmeric	16
Thymol	[89-83-8]	Thyme	14
Methyl salicylate	[119-36-8]	Wintergreen	4

flavor, which is described as "sweet, spicy, floral, with a fatty herbaceous undertone." Safranal (2,6,6-trimethyl-1,3-cyclohexadiene-1-carboxaldehyde) has been generally considered to be the character-impact compound of saffron; however, a recent investigation has also identified two other potent compounds, 4,4,6-trimethyl-2,5-cyclohexadien-1-one and an unknown constituent possessing "saffron, stale, and dried hay" aroma attributes [17]. Representative structures for spice impact compounds are shown in Figure 9.1.

A key component in both chili powder and curry powder, cumin is the dried seed of the herb *Cuminum cyminum*, a member of the parsley family. Cuminaldehyde is the principal contributor to the spice's aroma and flavor,

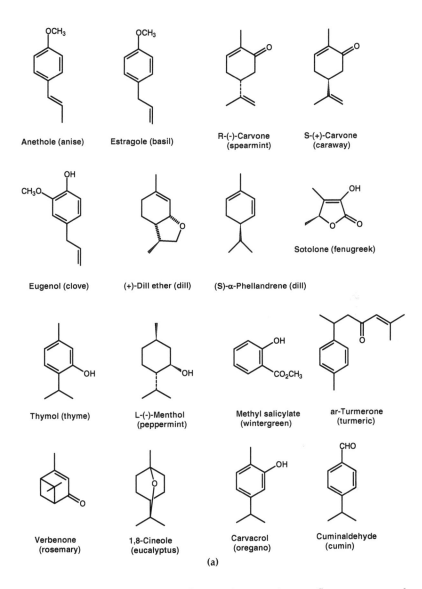

Figure 9.1(a) Representative herb and spice character-impact flavor compounds.

which imparts a strong musty or earthy character, with green grassy notes contributed by *p*-1,3 and 1,4-menthadienals. *trans*-2-Dodecenal, possessing a persistent fatty–citrus–herbaceous odor, is a character-impact component of coriander, along with *d*-linalool [18].

Thus far very little has been reported concerning the importance of sesquiterpenes in natural flavors. Although a high proportion of monoterpenes are present in black pepper oil with little consensus as to their relative importance [12], sesquiterpene isomers α-*trans*- and α-*cis*-bergamotene provide a distinct odor of ground black pepper [19a], whereas (+)-sabinine and linalool

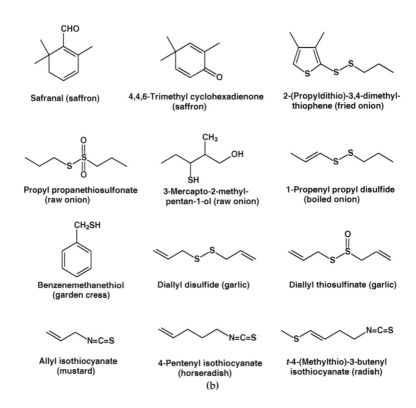

Figure 9.1(b) Representative herb and spice character-impact flavor compounds.

play a supporting role [19b]. It is well known that (+)-a-phellandrene, the main constituent of dill (*Anethum graveolus* L.) greatly contributes to the sensory impression of dill herb [12]. However, fresh dill character-impact appears to be a synergistic relationship contributed primarily by (+)-α-phellandrene with a modifying effect from "dill ether", 3,9-epoxy-*p*-menth-1-ene [20].

The Allium family includes garlic, onion, leek, and chive. All are comprised of sulfur-containing character-impact compounds. The aroma impact constituents of garlic are diallyl disulfide and the corresponding thiosulfinate derivative (allicin), which are enzymatically released from a sulfoxide flavor precursor (alliin) during the crushing of garlic cloves [21].

The flavor chemistry of sulfur compounds in onion is quite complex [22,23]. Early reports of polysulfides and thiosulfinates were later demonstrated to be thermal artifacts from gas chromatographic analysis [23]. Character-impact sulfur compounds have been proposed for fresh, boiled, and fried onion. In raw, fresh onion, propyl propanethiosulfinate, propenyl propanethiosulfinate thiopropanal S-oxide, and propyl methanethiosulfinate are impact contributors [21,22]. Several compounds contribute to the aroma character of cooked onion, of which dipropyl disulfide and allyl propyl disulfide provide key impact [21]. Fried onion aroma is formed by heating the latter compound, and is characterized by 2-(propyldithio)-3,4-dimethylthiophene, which has an odor threshold of 10 to 50 ng/L in water. A recent study identified a new potent aroma compound from raw onions, 3-mercapto-2-methylpentan-1-ol [24]. The flavor impact of this thiol is strongly dependent on concentration; at 0.5 ppb, it provides a pleasant brothlike, sweaty, onion, and leeklike flavor, whereas at high levels it provides a strong, unpleasant onionlike quality.

Isothiocyanates are character-impact constituents which provide pungency and typical flavor to mustard (allyl isothiocyanate), radish, *trans*-4-(methylthio)-3-butenyl- and *trans*-4-(methylthio)butyl isothiocyanate) and horseradish (4-pentenyl- and 2-phenylethyl isothiocyanate) [21]. Benzenemethanethiol was identified as the character-impact flavor for the unique flavor of garden cress (*Lepidium sativum*) seed, and also occurs in the volatile extracts of potatoes [25]. Garden cress (peppergrass) is classified in the *Brassica* (*Cruciferae* or mustard) family, and its leaves and seeds were historically used as salad greens and as a spicy condiment because of their "peppery" flavor.

B.   Fruit flavors

Aroma constituents of the essential oils from fruits such as lime, lemon, and orange were among the first character-impact compounds identified by flavor chemists. Fruit flavors are a subtle blend of characterizing volatile compounds, supported by fruit sugars, organic acids, and noncharacterizing volatile esters. Fruit aromatics tend to be present in concentrations of greater abundance (< 30 ppm) than other foods, which facilitated early analytical studies. The volatile composition of fruits is extremely complex, and noncharacterizing flavor esters are common across species. A compilation of character-impact compounds found in fruit flavors is summarized in Table 9.2.

The combination of ethyl 2-methyl butyrate, β-damascenone, and hexanal contributes to the characteristic flavor of the Delicious apple [26,27]. This blend of character-impact flavors combines "apple ester" and "green apple" notes, which fluctuate with apple ripeness and seasonality. β-Damascenone is an unusually potent aroma compound with a threshold of 2 pg/g in water, and it also occurs in natural grape and tomato flavors [27].

Two important character-impact compounds of strawberry flavor, are the furanones 2,5-dimethyl-4-hydroxy-2*H*-furan-3-one (Furaneol™) and 2,5-dimethyl-4-methoxy-2*H*-furan-3-one (mesifuran) [28]. However, at various concentrations Furaneol™ can simulate other flavors, e.g., pineapple [14] or Muscadine grape [29] at low levels and caramel at high levels. Mesifuran exhibits a sherrylike aroma, and is a contributor to sherry and French white wine aroma. Other important character-impact compounds of strawberry flavor are methyl cinnamate and ethyl 3-methyl-3-phenylglycidate, a synthetic aroma chemical [4,30]. Representative chemical structures for fruit flavor impact compounds are shown in Figure 9.2.

The character-impact component for Concord (*Vitis labrusca*) grape has long been known as methyl anthranilate. Subsequently, ethyl 3-mercaptopropionate was identified in Concord grape, and it possesses a pleasant fruity fresh grape aroma at low parts-per-million levels [31]. 2-Aminoacetophenone and mesifuran are also significant contributors in Concord to its

Table 9.2 Character-Impact Flavor Compounds in Fruits

Character-Impact Compounds	CAS Registry No.	Occurrence	Reference
Ethyl-2-methyl butyrate	[7452-79-1]	Apple	4
β-Damascenone	[23696-85-7]	Apple	4,27
iso-Amyl acetate	[123-92-2]	Banana	4
4-Methoxy-2-methyl-2-butanethiol	[94087-83-9]	Blackcurrant	21
iso-Butyl 2-butenoate	[589-66-2]	Blueberry	4
Benzaldehyde	[100-52-7]	Cherry	4
p-Tolyl aldehyde	[1334-78-7]	Cherry	4
Ethyl heptanoate	[106-30-9]	Cognac brandy	4
Methyl anthranilate	[134-20-3]	Grape, Concord	21
Ethyl-3-mercaptopropionate	[5466-06-8]	Grape, Concord	31
Linalool	[78-70-6]	Grape, Muscat	34
4-Mercapto-4-methyl-2-pentanone	[19872-52-7]	Grape, Sauvignon	5
		Grapefruit	38
Nootkatone	[4674-50-4]	Grapefruit	4
1-*p*-Menthene-8-thiol	[71159-90-5]	Grapefruit	41
Citronellyl acetate	[150-84-5]	Kumquat	44
Citral (neral + geranial)	[5392-40-5]	Lemon	4
α-Terpineol	[98-55-5]	Lime	4
Citral (neral + geranial)	[5392-40-5]	Lime	4
2,6-Dimethyl-5-heptenal	[106-72-9]	Melon	14
Z-6-Nonenal	[2277-19-2]	Melon	26
Methyl N-methylanthranilate	[85-91-6]	Orange, Mandarin	43
Thymol	[89-83-8]	Orange, Mandarin	43
3-Methylthio-1-hexanol	[5155-66-9]	Passion fruit	21
2-Methyl-4-propyl-1,3-oxathiane	[67715-80-4]	Passion fruit	21
γ-Undecalactone	[104-67-6]	Peach	4
6-Pentyl-2H-pyran-2-one	[27593-23-3]	Peach	4
Ethyl trans-2,cis-4-decadienoate	[3025-30-7]	Pear, Bartlett	4
Allyl caproate	[123-68-2]	Pineapple	14
Ethyl 3-(methylthio)propionate	[13327-56-5]	Pineapple	49
Allyl 3-cyclohexylpropionate	[2705-87-5]	Pineapple	14
4-(*p*-Hydroxyphenyl)-2-butanone	[5471-51-2]	Raspberry	4,45,46
trans-α-ionone	[127-41-3]	Raspberry	4,45,46
Ethyl 3-methyl-3-phenylglycidate	[77-83-8]	Strawberry	4
Ethyl-2-methyl butyrate	[7452-79-1]	Apple	4
Furaneol	[3658-77-3]	Strawberry; Muscat	28,29
Mesifuran	[4077-47-8]	Strawberry; Sherry	28
(Z,Z)-3,6-Nonadienol	[53046-97-2]	Watermelon	26

"foxy" and "candylike" notes, respectively. Ethyl heptanoate elicits an odor reminiscent of cognac brandy [4]. These aroma characters are in significant contrast to those resulting from grape varieties (*Vitis vinifera*) used for production of table wines, whose flavors derive from the specific variety of grape, vinification, maturation, and aging conditions. It is well-established that characteristic wine flavors are produced from secondary metabolites of

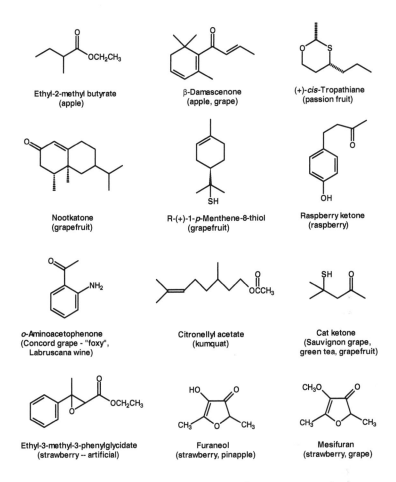

Figure 9.2 Representative fruit character-impact flavor compounds.

grapes and yeast during the fermentation process, which increase the chemical and aroma complexity of wine [32]. For example, in Pinot Noir wines, there is no single compound, but rather a complex blend of odorants including 2-phenylethanol, ethyl butyrate, 3-methylbutyl alcohol/acetate and benzaldehyde, among others [33], that characterizes its fruity "plum, cherry, ripe blackberry" aroma. However, other wines made from certain *vinifera* grape varieties possess unique flavor impact compounds, including: "flowery, muscat" character from linalool in Muscat wine, "green pepper, earthy" from 3-isopropyl- and 3-isobutyl-2-methoxy pyrazines in Sauvignon blanc wine, "foxy" from 2-aminoacetophenone in Labruscana wine, "floral, lychee" from cis-rose oxide in Gewurztraminer, and "sulfury" character from 3-mercaptohexan-1-ol in Grenache, Merlot, and Cabernet wines ([34] and references cited therein). Red Tannat wines, produced in southern Uruguay, frequently described as "spicy, mintlike," contain 1,8-cineole (eucalyptol) as a flavor character. In Tannat fermentations, 1,8-cineole is produced by chemical rearrangements of limonene or α-terpineol, and levels can be increased by heating to 45°C at wine pH (3.2) [35]. As previously discussed, 4-mercapto-4-methyl-2-pentanone (cat ketone) provides "catty" character in Sauvignon blanc (and also Scheurebe) wine at an aroma perception threshold below 3 ppt [5,36]. It is also a characteristic flavorant in Japanese green tea (*sen-cha*) [37], and contributes significantly to the flavor of hand-squeezed grapefruit juice [38].

The mechanism of thioketone formation involves an aldol condensation between two molecules of acetone to yield 4-hydroxy-4-methyl-2-pentanone **1** as shown in Figure 9.3. Dehydration of this intermediate forms 4-methyl-3-penten-2-one (mesityl oxide, **2**). Hydrogen sulfide (from degradation of cysteine and methionine amino acids in proteins) can readily undergo a Michael-type addition to the enone moiety of mesityl oxide to generate the thioketone **3** [39].

Blackcurrant flavor is very popular in Europe, and is associated with numerous health-related functional foods and alcoholic drinks (cassis liqueur). The key aroma component in blackcurrant is 2-methoxy-4-methyl-4-butanethiol [21]. The "catty or ribes" flavor of blackcurrant (*Ribes nigrum*) was earlier attributed to "cat ketone", but it was recently shown to be absent during flavor and sensory analysis of blackcurrant juice concentrates [40].

Two character-impact compounds have been proposed for grapefruit flavor, the first being nootkatone, a sesquiterpene. The fresh juicy note of grapefruit juice is attributable to 1-*p*-menthene-8-thiol. This compound has a detection threshold of 10^{-1} ppt, among the lowest values reported for aroma chemicals [41]. The (+)-R-isomer was found to have a lower aroma threshold in water than the racemic mixture, and it imparts a pleasant, fresh grapefruit juice character, as opposed to the extremely obnoxious sulfur note contributed by the (−)-S-epimer.

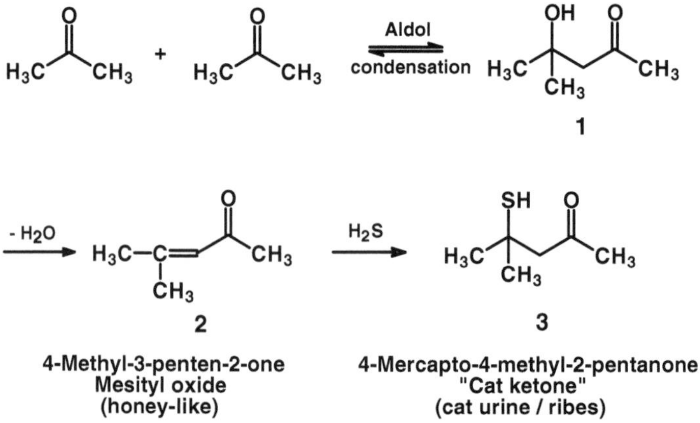

Figure 9.3 Mechanism for formation of 4-mercapto-4-methyl-2-pentanone ("cat ketone").

Among other citrus flavors, the basic flavor-impact compound of lemon is citral, a mixture of neral and geranial isomers, which together comprise the aroma impression. The flavor character of lime results from a combination of α-terpineol and citral, even though limonene is the most abundant, but sensorially immaterial, volatile in lime and other citrus oils. In contrast to other citrus flavors, orange flavor lacks a specific character-impact compound, and the current belief is that orange flavor is the result of a complex mixture of C_9–C_{13} aldehyde, terpene, and norisoprenoid volatiles blended in specific proportions. [42]. For Mandarin and tangerine flavors, methyl-N-methyl anthranilate and thymol are associated with character impact, with additional contributions from β-pinene and γ-terpinene [43].

Kumquat (*Fortunella japonica* Swingle) is another fruit in the citrus genus that contains characteristic terpene compounds found in other citrus oils. The flesh is quite sour, and the peel is often consumed with the flesh. Two sensorially significant diterpene esters, citronellyl formate and citronellyl acetate, were identified in kumquat peel oil by organoleptic evaluation with GC-O. However, citronellyl acetate was found to possess the sensory character that most contributes the aroma of kumquat [44].

Lactones have characteristic aromas that are attributable to peach, coconut, and dairy flavors and occur in a wide variety of foods. The γ-lactones, specifically γ-undecalactone and lesser for γ-decalactone, possess intense peachlike odors [14]. A doubly unsaturated δ-decalactone, 6-pentyl-2H-pyran-2-one, also has an intense peach character [4]. As a point of distinction the C_{10}–C_{12} δ-lactones, particularly the "creamy-coconut" note of δ-decalactone, are flavor constituents of coconut as well as cheese and dairy products [14]. Wine lactone (3a,4,5,7a-trtrahydro-3,6-dimetylbenzofuran-2(3H)-one), which conveys an intense "coconut-like, woody, sweet" aroma, has been reported as a background note in Gewurtztraminer and Scheurebe white wines [36].

In addition to the character-impact compound of raspberry, 4-(4-hydroxyphenyl)-butan-2-one (raspberry ketone), alpha- and beta-ionone, geraniol, linalool, Furaneol, maple furanone, and other furaneols were concluded to be of importance to raspberry aroma [45,46]. The odor threshold of raspberry ketone was measured at 1 to 10 µg/kg. The ionones have chemical structure similarities and potencies compared to β-damascenone [14].

The aroma profiles of Evergreen, Marion, and Chickasaw blackberries are complex, as no single volatile was unanimously described as characteristically "blackberry" [47,48]. Flavors of importance include ethyl butyrate, Furaneol, and other furanones, linalool, and β-damascenone. Blackberry aroma character was shown to differ for the same variety, depending on growing region [48].

The "tropical" category is one of the most important areas for new discoveries of key impact flavor compounds. Analyses of passion fruit and durian flavors have produced identifications of many potent sulfur aroma compounds [21]. Among these is tropathiane, 2-methyl-4-propyl-1,3-oxathiane, which has an odor threshold of 3 ppb [18]. For pineapple, 2-propenyl hexanoate (allyl caproate) exhibits a typical pineapple character [14], however Furaneol, ethyl 3-methylthiopropionate, and ethyl-2-methylbutyrate are important supporting character-impact compounds [49]. The latter ester contributes the background "apple" note to pineapple flavor. Another character-impact compound, allyl 3-cyclohexylpropionate, has not been discovered in nature, but it provides a sweet, fruity pineapple flavor note [14].

Characterizing flavors for melons include (Z)-6-nonenal, which contributes a typical melon aroma impression, and (Z,Z)-3,6-nonadienol for watermelon rind aroma impact [26]. 2,6-Dimethyl-5-hepten-1-al (Melonal™) has not been identified in melon, but provides a melonlike note in compounded flavors [14]. The flavor of muskmelons is more complex, with methyl 2-methylbutanoate, Z-3-hexenal, E-2-hexenal, and ethyl 2-methylpropanoate identified as the primary, noncharacterizing odorants [50].

Although an important constituent to both varieties, the flavor of sweet cherries (*Prunus avium*) is less dominated by the character-impact compound, benzaldehyde, than is the flavor profile of sour cherries (*Prunus cerasus*) [51].

C. Vegetable flavors

Recent aroma research has been devoted to the identification of key flavor compounds in vegetables and is the subject of several contemporary reviews [49,52,53]. Cucumbers, sweet corn, and tomatoes are botanically classified as fruits; however, for flavor considerations they are regarded as vegetables as they are typically consumed with the savory portion of the meal. Overall, the knowledge base of character-impact compounds for vegetables is much smaller than other flavor categories, and warrants further investigation.

Identifying flavor-impact compounds in vegetables depends considerably on how they are prepared (cutting, blending), and the form in which they are consumed (raw vs. cooked). For example, the character impact of fresh tomato is delineated by 2-*iso*-butylthiazole and (Z)-3-hexenal, with modifying effects from β-ionone and β-damascenone [49]. Alternatively, dimethyl sulfide is a major contributor to the flavor of thermally processed tomato paste [49,52]. Dimethyl sulfide is also a flavor impact compound for both canned cream and fresh corn, while 2-acetyl-1-pyrroline provides a "corn chip" character. Hydrogen sulfide, methanethiol, and ethanethiol may further contribute to the aroma of sweet corn owing to their low odor thresholds [53]. A summary of character-impact compounds for vegetable flavors is outlined in Table 9.3.

The character-impact compound of green bell pepper, 2-isobutyl-3-methoxypyrazine, was the first example of a high-impact aroma compound because of its exceptionally low odor threshold of 2 ppt [18]. A similarly low-odor-threshold compound, geosmin, is the character impact of red beets and is detectable at a 100 ppt concentration. The flavor of raw peas and peapods is attributable to 2-*iso*-propyl-3-methoxypyrazine [52]. The importance of C_9 aldehydes to the character impact of cucumber flavor was recently confirmed by calculating their odor unit values (ratio of concentration to

Table 9.3 Character-Impact Flavor Compounds in Vegetables

Character-Impact Compounds	CAS Registry No.	Occurrence	Reference
Dimethyl sulfide	[75-18-3]	Asparagus	4
1,2-Dithiacyclopentene	[288-26-6]	Asparagus, heated	52
Geosmin	[19700-21-1]	Red beet	52
4-Methylthiobutyl isothiocyanate	[4430-36-8]	Broccoli	53
Dimethyl sulfide	[75-18-3]	Cabbage	52
Methyl methanethiosulfinate	[13882-12-7]	Sauerkraut	56
2-sec-Butyl-3-methoxypyrazine	[24168-70-5]	Carrot (raw)	52
3-(MeS)propyl isothiocyanate	[505-79-3]	Cauliflower	53
3-Butylphthalide	[6066-49-5]	Celery	53
Sedanolide	[6415-59-4]	Celery	53
(Z)-3-Hexenyl pyruvate	[68133-76-6]	Celery	4
Dimethyl sulfide	[75-18-3]	Corn	53
2-Acetyl-2-thiazoline	[29926-41-8]	Corn (fresh)	53
(E,Z)-2,6-Nonadienal	[557-48-2]	Cucumber	50
(E)-2-Nonenal	[2463-53-8]	Cucumber	50
2-iso-Butyl-3-methoxypyrazine	[24683-00-9]	Green bell pepper	52
1-Octen-3-ol	[3391-86-4]	Mushroom	52
1-Octen-3-one	[4312-99-6]	Mushroom	52
p-Mentha-1,3,8-triene	[18368-95-1]	Parsley	58
2-iso-Propyl-3-methoxypyrazine	[25773-40-4]	Pea (raw)	52
3-Methylthiopropanal	[3268-49-3]	Potato (boiled)	21
2-iso-Propyl-3-methoxypyrazine	[25773-40-4]	Potato (earthy)	52
2-Ethyl-6-vinylpyrazine	[32736-90-6]	Potato (baked)	52
(Z)-3-Hexenal	[6789-80-6]	Tomatillo	54
(E,E)-2,4-Decadienal	[25152-84-5]	Tomatillo	54
2-iso-Butylthiazole	[18640-74-9]	Tomato (fresh)	21
(Z)-3-Hexenal	[6789-80-6]	Tomato (fresh)	49

odor threshold). (E,Z)-2,6-Nonadienal and (Z)-2-nonenal were determined to be the principal odorants of cucumbers [50].

Tomatillo (*Physalis ixocarpa* Brot.) is a solanaceous fruit vegetable similar in appearance to a small green tomato, which is used to prepare green salsas in various Mexican dishes. The character-impact compounds in tomatillo were recently established as (Z)-3-hexenal, (E,E)-2,4-decadienal, and nonanal, which impart its dominant "green" flavor [54]. Similar to tomato flavor, β-ionone and β-damascenone provide modifying effects; however, tomatillo does not contain 2-iso-butylthiazole, a key character-impact compound of tomato.

Among the Cruciform vegetables, cooked cabbage owes its dominant character-impact flavor to dimethyl sulfide. In raw cabbage flavor, allyl isothiocyanate contributes sharp, pungent horseradish-like notes [53]. Methyl methanethiosulfinate was observed to provide the character impact of sauerkraut flavor, and occurs in Brussels sprouts and cabbage [55,56]. Compounds likely to be important to the flavor of cooked broccoli include dimethyl sulfide and trisulfide, nonanal, and erucin (4-(methylthio)butyl isothiocyanate) [52]. Cooked cauliflower contains similar flavor components as broccoli, with the exception that 3-(methylthio)propyl isothiocyanate is the characterizing thiocyanate [52]. Representative structures for vegetable flavor impact compounds are presented in Figure 9.4.

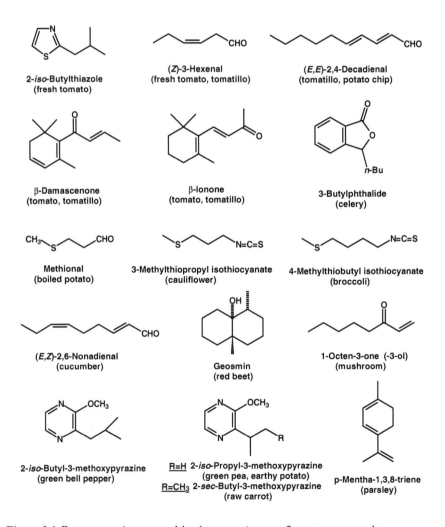

Figure 9.4 Representative vegetable character-impact flavor compounds.

Potato flavor is greatly influenced by methods of cooking or preparation. Raw potato contains the characteristic "earthy aroma" component, 2-*iso*-propyl-3-methoxypyrazine. A character-impact compound common to boiled and baked potatoes is methional (3-(methylthio)propanal). Baked potatoes contain Maillard products such as 2-ethyl-3-methylpyrazine (earthy and nutty) and 2-ethyl-6-vinylpyrazine (buttery and baked potato) [52]. In potato chips and French-fried potatoes, the potato flavor character of methional is modified by volatile aromatics from frying oils, such as (*E,E*)-2,4-decadienal, and thermally-generated alkyl oxazoles possessing lactone-like flavors [52,57].

Aroma-impact compounds which are universal to all varieties of mushrooms include 1-octen-3-ol (1 ppb threshold) and 1-octen-3-one (0.05 ppb threshold), both which have been described as having a fresh, wild mushroom aroma [18]. However, 1-octen-3-one also possesses a metallic odor, particularly in the context of oxidized oils [52].

The characteristic compound for raw carrot is 2-*sec*-butyl-3-methoxypyrazine, which has an extremely low (2 ppt) threshold value. Its sniffing-port aroma in GC-O has been described as "raw carroty" [52]. Unsaturated aldehydes contribute to the flavor of cooked carrot, the most significant being (*E*)-2-nonenal (fatty-waxy) [52]. Compounds important to the aroma of celery include two lactones, 3-butylphthalide, and sedanolide (3-butyl-3*a*,4,5,6-tetrahydrophthalide), which provide impact at low concentration (< 3 ppm). Other terpene hydrocarbons such as β-selinene and limonene are present in greater abundance, but do not significantly contribute to celery flavor. A recent assessment of its odor threshold by Takeoka suggests that 3-butylphthalide is the most significant character-impact compound for celery [53].

Key parsley aroma compounds were recently identified [58]. The primary flavor contributors were found to include *p*-mentha-1,3,8-triene (terpeny and parsley-like), myrcene (metallic and herbaceous), 2-*sec*-butyl-3-methoxypyrazine (musty and earthy), myristicin (spicy), linalool (coriander), (*Z*)-6-decenal (green and cucumber), and (*Z*)-3-hexenal (green).

D. *Maillard-type, brown, and cereal flavors*

The flavor characteristics of heated sugar compounds that possess caramel, burnt sugar, and maple notes include a family of structures that contain a methyl enol-one group [59]. As previously discussed, Furaneol (2,5-dimethyl-4-hydroxy-(2*H*)-furan-3-one) has a sweet caramel, burnt-sugar flavor with appreciable fruitiness, and occurs in beer, Arabica coffee, and white bread crust [60]. Maltol (3-hydroxy-2-methyl-4*H*-pyran-4-one) exhibits a sweet, burnt sugar, caramel note similar to Furaneol, but not as strong or fruity. Ethyl maltol does not occur in nature; however, it possesses a very intense sweet, caramel-like odor, which is four to six times more potent than maltol [14]. Cyclotene (3-methyl-2-cyclopenten-2-ol-1-one) has a strong maple–caramel flavor. 2-Methyltetrahydrofuran-3-one has a very pleasant, sweet-caramel character [18]. Sotolon (4,5-dimethyl-3-hydroxy-2(5*H*)-furanone) was identified as a potent flavor-impact compound from raw cane sugar. At low concentrations, its aroma character is caramel-like, and elicits a powerful caramel, sweet, burnt-sugar note, typical of unrefined cane sugar [61]. A summary of character-impact compounds for thermally generated flavors is outlined in Table 9.4.

Table 9.4 Character-Impact Flavor Compounds in Cooked Flavors and Maillard-type Systems (Chocolate, Coffee, and Caramelized Sugar)

Character-Impact Compounds	CAS Registry No.	Occurrence	Reference
5-Methyl-2-phenyl-2-hexenal	[21834-92-4]	Chocolate	81
2-Methoxy-5-methylpyrazine	[2882-22-6]	Chocolate	4,82
iso-Amyl phenylacetate	[102-19-2]	Chocolate	4,82
2-Furfurylthiol	[98-02-2]	Coffee	62
		Wine (barrel-aged)	65
Furfuryl methyl disulfide	[57500-00-2]	Coffee (mocha)	18
2-Methyl-3-furanthiol	[28588-74-1]	Wine (barrel-aged)	65
Furaneol	[3658-77-3]	Fruity, burnt sugar	60
Sotolon	[28664-35-9]	Brown sugar	61
Phenylacetaldehyde	[122-78-1]	Honey	82
3-Methylbutanal	[590-86-3]	Malt	70
Maltol	[118-71-8]	Cotton candy	14
Ethyl maltol	[4940-11-8]	Caramel, sweet	14
2-Hydroxy-3-methyl-2-cyclopenten-1-one	[80-71-7]	Maple	14
2-Acetyl-1,4,5,6-tetrahydropyridine	[25343-57-1]	Cracker (saltine)	69
(*E,E,Z*)-2,4,6-Nonatrienal	[100113-52-8]	Oat flake	77
2-Acetyl-1,4,5,6-tetrahydropyridine	[25343-57-1]	Popcorn	67,68
2-Acetyl-1-pyrroline	[85213-22-5]	Popcorn	67,68
2-Propionyl-1-pyrroline	[133447-37-7]	Popcorn	67
2-Acetylpyrazine	[22047-25-2]	Popcorn	67
2-Acetyl-2-thiazoline	[22926-41-8]	Roasty, popcorn	79
5-Acetyl-2,3-dihydro-1,4-thiazine	[164524-93-0]	Popcorn (model sys.)	71
3-Thiazolidineethanethiol	[317803-03-5]	Popcorn (model sys.)	72
2-Acetyl-1-pyrroline	[85213-22-5]	Wheat bread crust	68,70
		Rice (Basmati)	68,70
2-Ethyl-3,5-dimethylpyrazine	[13925-07-0]	Potato chip	79
2,3-Diethyl-5-methylpyrazine	[18138-04-0]	Potato chip	79
2-Vinylpyrazine	[4177-16-6]	Roasted potato	79
(Z)-2-propenyl-3,5-dimethylpyrazine	[55138-74-4]	Roasted potato	79
Guaiacol	[90-05-1]	Smoky	4
4-Vinylguaiacol	[7786-61-0]	Smoky	18

An alternative source of characteristic heated flavor compounds arises via the Maillard pathway, the thermally induced reaction between amino acids and reducing sugars. Aroma constituents in chocolate, coffee, toasted cereal grains, wheat bread crust, and popcorn are products of Maillard reactions, in addition to flavors in roasted nuts and meats, which are discussed in Section 9.II.E and Section 9.II.F, respectively. Guaiacols occur as pyrolysis products of carbohydrates or lipids in smoked and char-broiled meats.

2-Furfurylthiol is the primary character-impact compound for the aroma of roasted Arabica coffee [62]. It has a threshold of 5 ppt and smells like freshly brewed coffee at concentrations between 0.01 and 0.5 ppb [63]. At higher concentrations it exhibits a stale coffee, sulfury note. Other potent odorants in roasted coffee include 5-methylfurfurylthiol (0.05 ppb threshold), which smells meaty at 0.5 to 1 ppb, and changes character to a sulfury mercaptan note at higher levels [63]. Furfuryl methyl disulfide has a sweet mocha coffee aroma [18]. Although other sulfur compounds such as 3-mercapto-3-methylbutyl formate and 3-methyl-2-buten-1-thiol were previously thought to be important factors for coffee aroma, recent studies have confirmed that the flavor profile of brewed coffee is primarily contributed by 2-furfurylthiol, 4-vinylguaiacol, and "malty"-smelling Strecker aldehydes, among others [62,64]. For wines and champagnes aged in toasted oak barrels, 2-furfurylthiol and 2-methyl-3-furanthiol have been identified as providing their "roasted-coffee" and "cooked meat" aroma characters, respectively [65].

Cereal grains, including wheat, rice, corn and oats, have characteristic "toasted, nutty" flavors (pyrazines, tetrahydropyridines, pyrrolines, etc.) that are thermally generated through Maillard pathways [66]. The principal aroma-impact compounds of freshly prepared popcorn were determined by Schieberle as 2-acetyltetrahydropyridine, 2-acetyl-1-pyrroline, and 2-propionyl-1-pyrroline [67,68]. The cracker-like aroma of the tetrahydropyridine, which exists in two tautomeric forms (2-acetyl-1,4,5,6- and 2-acetyl-3,4,5,6-tetrahydropyridine), was previously identified as the character compound of saltine crackers [69]. The decrease of these compounds during storage was directly correlated with staling flavor. Another "popcorn-like" odorant, 2-acetylpyrazine, was determined to have minor contribution to the aroma of fresh popcorn, because of its considerably lower odor activity [70]. Two novel, highly intense "roasty, popcorn-like" aroma compounds were recently identified in Maillard model systems: 5-acetyl-2,3-dihydro-1,4-thiazine (0.06 ppt odor threshold) from reaction of ribose with cysteine [71]; and N-(2-mercaptoethyl)-1,3-thiazolidine (3-thiazolidineethanethiol) (0.005-ppt odor threshold) from the reaction of fructose with cysteamine [72]. Neither of these characteristic "popcorn" flavor compounds have been reported in food aromas to date. Representative chemical structures for thermally-generated flavor impact compounds are shown in Figure 9.5.

The flavor formed by the cooking of fragrant rice (e.g., basmati) is described as popcorn-like; hence, it is not surprising that 2-acetyl-1-pyrroline is the character-impact volatile [70]. In masa corn tortillas, 2-aminoacetophenone

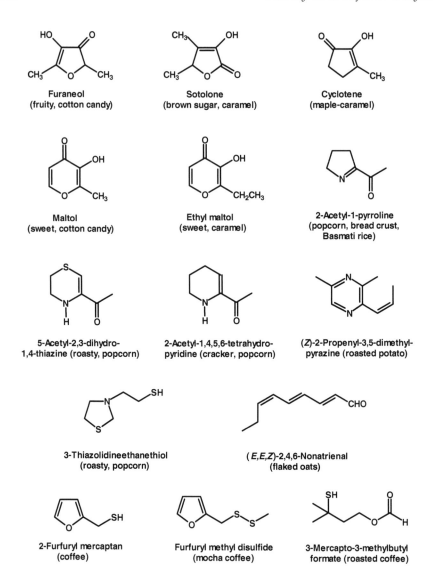

Figure 9.5 Representative Maillard-generated and cereal-like character-impact flavor compounds.

provided the character impact resulting from the lime treatment of corn [73], whereas in corn chips its contribution is modified by 2-acetyl-1-pyrroline and unsaturated aldehydes [74].

In addition to Maillard pathways, "cereal-like" flavor character is provided by trace volatiles formed through lipoxygenase oxidation of unsaturated lipids, such as linoleic and linolenic acids. One of the first studies on oat flavor by Heydanek and McGorrin established the importance of C_8–C_{10}

unsaturated aldehydes and ketones as key contributors to the flavor of dried oat groats [75,76]. Recently, Schuh and Schieberle identified the character-impact compound (E,E,Z)-2,4,6-nonatrienal as eliciting the typical "cereal, sweet" aroma of rolled oat flakes [77]. This odorant exhibits an extremely low odor threshold (0.0002 ppt), among the lowest reported to date.

Two compounds that create the characteristic odor notes in the pleasant aroma of wheat bread crust have been identified as the "popcorn-like" 2-acetyl-1-pyrroline and 2-acetyltetrahydropyridine [70]. The aroma of the bread crumb portion is principally due to lipid-derived unsaturated aldehydes such as (E)-2-nonenal and (E,E)-2,4-decadienal, which create stale aromas at high levels. The malty notes which predominate in yeast and sourdough breads are attributed to 2- and 3-methylbutanal and Furaneol [70,78].

"Potato chip" aroma is associated with the pyrazines 2-ethyl-3,5-dimethylpyrazine and 2,3-diethyl-5-methylpyrazine, whereas 2-vinylpyrazine and (Z)-2-propenyl-3,5-dimethylpyrazine provide an intense "roasted potato" odor [79]. A combination of methional, 2-acetyl-1-pyrroline, phenylacetaldeyde, and butanal are important flavor characters for extruded potato snacks [25].

Cocoa and chocolate represent a highly complex flavor system for which no single character impact has been identified. Vanillin and Furaneol contribute to the sweet, caramel background character of milk chocolate, whereas 3-methylbutanal provides its "malty" flavor [80]. 5-Methyl-2-phenyl-2-hexenal provides a "deep bitter, cocoa" note, and is the aldol reaction product from phenylacetaldehyde and 3-methylbutanal, two Strecker aldehydes formed in chocolate [81]. 2-Methoxy-5-methylpyrazine and iso-amyl phenylacetate have a "chocolate, cocoa, nutty" and "cocoalike" notes, respectively, and both are used in synthetic chocolate flavors [82]. Systematic studies of key odorants in milk chocolate were performed using aroma extract dilution analysis; however, character-impact compounds unique to chocolate flavor were not reported [80,83].

E. Nut flavors

Pyrazines are the major compound classes in peanuts, formed through the thermally induced Maillard reaction (with the exception of methoxy pyrazines) [84]. Two pyrazines that represent peanut flavor character are 2,5-dimethylpyrazine (nutty) and 2-methoxy-5-methylpyrazine (roasted nutty) (Table 9.5).

Benzaldehyde has long been known as the character impact of oil of bitter almond. It possesses an intense almondlike flavor in the context of savory applications; in sweet systems, it becomes cherrylike. 5-Methyl-2-thiophene-carboxaldehyde also provides almond flavor character and occurs naturally in roasted peanuts [82].

The character-impact compound of hazelnuts, (E)-5-methyl-2-hepten-4-one (filbertone), undergoes isomerization during the roasting process [85]. Of the four possible geometric and enantiomeric isomers formed, all exhibit the

Table 9.5 Character-Impact Flavor Compounds in Nuts

Character-Impact Compounds	CAS Registry No.	Occurrence	Reference
Benzaldehyde	[100-51-6]	Almond	4
5-Methyl-2-thiophenecarboxal	[13679-70-4]	Almond	4,82
γ-Nonalactone	[104-61-0]	Coconut	84
δ-Decalactone	[705-86-2]	Coconut	84
Methyl(methylthio)pyrazine	[21948-70-9]	Hazelnut	82
(E)-5-Methyl-2-hepten-4-one	[81925-81-7]	Hazelnut	85
2,5-Dimethylpyrazine	[123-32-0]	Peanut	4
2-Methoxy-5-methylpyrazine	[68358-13-5]	Peanut	4

typical hazelnut aroma, but the *trans*-(S)-isomer has the strongest impact. Methyl(methylthio)pyrazine is a synthetic aroma chemical with the character of roasted almonds and hazelnuts [82]. Structures are presented in Figure 9.6.

As previously discussed, the δ-lactones (e.g., δ-decalactone and δ-octalactone) possess a coconut flavor character. However, γ–nonalactone has the most intense coconut-like aroma as an individual character-impact compound, but it occurs only in artificial coconut flavors [84]. As the side-chain length increases, the character of γ-lactones changes to peachlike [14].

F. Meat and seafood flavors

Sulfur-containing heterocyclic compounds are associated with meaty characteristics. Two compounds with the most potent meaty impact include 2-methyl-3-furanthiol (1 ppt) and the corresponding dimer,

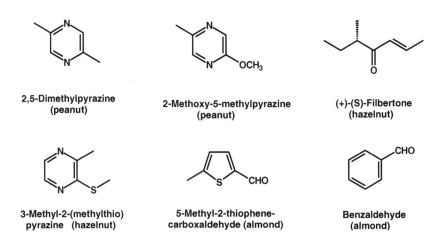

2,5-Dimethylpyrazine (peanut)

2-Methoxy-5-methylpyrazine (peanut)

(+)-(S)-Filbertone (hazelnut)

3-Methyl-2-(methylthio) pyrazine (hazelnut)

5-Methyl-2-thiophene-carboxaldehyde (almond)

Benzaldehyde (almond)

Figure 9.6 Representative character-impact flavor compounds in peanuts and tree nuts.

bis-(2-methyl-3-furyl) disulfide (0.02 ppt) [21]. Both substances have been identified in cooked beef and chicken broth, and have a strong meaty quality upon dilution. 2-Methyl-3-furanthiol also occurs in canned tuna fish aroma [86]. The disulfide has a recognizable aroma character of "rich aged-beef, prime rib" [18]. Interestingly, both compounds are produced from the thermal degradation of thiamin [87]. A related compound, 2-methyl-3-(methylthio)furan, is the character-impact compound for roast beef [21]. Other potent modifiers, such as 2-acetyl-2-thiazoline, impart a potent "roasty, popcorn" note, which enhances the meaty and roast flavor [88]. 2-Ethyl-3,5-dimethylpyrazine and 2,3-diethyl-5-methylpyrazine also contribute potent "roasty" notes to roast beef flavor [89]. A summary of character-impact compounds for meat and seafood flavors is shown in Table 9.6.

A brothy compound associated with boiled beef, 4-methylthiazole-5-ethanol (sulfurol), is a "reaction flavor" product from hydrolysis of vegetable protein. It is suspected that a trace impurity (2-methyltetrahydrofuran-3-thiol) in sulfurol is the actual "beef broth" character-impact compound [18]. Another reaction product flavor chemical, 2,5-dimethyl-1,4-dithiane-2,5-diol (the dimer of mercaptopropanone), has an intense chicken broth odor. A synthetic pyrazine, 2-pyrazineethanethiol, provides excellent pork character [18].

Table 9.6 Character-Impact Flavor Compounds in Meats and Fish

Character-Impact Compounds	CAS Registry No.	Occurrence	Reference
Dimethyl sulfide	[75-18-3]	Clam, oyster	93
4-Methylnonanoic acid	[45019-28-1]	Lamb	87
4-Methyloctanoic acid	[54947-74-9]	Lamb	87
2-Pentylpyridine	[2294-76-0]	Lamb	87
2-Acetyl-2-thiazoline	[22926-41-8]	Roasty (beef)	88
2-Methyl-3-(methylthio)furan	[63012-97-5]	Roast beef	21
4-Me-5-(2-hydroxyethyl)thiazole	[137-00-8]	Roasted meat	18
2-Ethyl-3,5-dimethylpyrazine	[13925-07-0]	Roasty (beef)	89
2,3-Diethyl-5-methylpyrazine	[18138-04-0]	Roasty (beef)	89
2-Methyltetrahydrofuran-3-thiol	[57124-87-5]	Brothy, meaty	18
2-Methyl-3-furanthiol	[28588-74-1]	Meat, beef Canned tuna fish	21
Bis-(2-methyl-3-furyl) disulfide	[28588-75-2]	Aged, prime-rib	21
12-Methyltridecanal	[75853-49-5]	Beef, stewed	91
(*E,E*)-2,4-Decadienal	[25152-84-5]	Chicken fat	18
2,5-Dimethyl-1,4-dithiane-2,5-diol	[55704-78-4]	Chicken broth	18
2-Pyrazineethanethiol	[35250-53-4]	Pork	18
(*Z*)-1,5-Octadien-3-one	[65767-22-8]	Salmon, cod	96
(*E,Z*)-2,6-Nonadienal	[557-48-2]	Trout, boiled	97
Pyrrolidino-2,4-(Me$_2$)dithiazine	[116505-60-3]	Roasted shellfish	99
5,8,11-Tetradecatrien-2-one	[85421-52-9]	Shrimp, cooked	98
2,4,6-Tribromophenol	[118-79-6]	Shrimp, ocean fish	100

Lipid components associated with meat fat, especially unsaturated alde-hydes, play a significant role in species characterization flavors. For example, (*E,Z*)-2,4-decadienal exhibits the character impact of chicken fat and freshly boiled chicken [90]. (*E,E*)-2,6-Nonadienal has been suggested as the compo-nent responsible for the tallowy flavor in beef and mutton fat [87]. 12-Meth-yltridecanal was identified as a species-specific odorant of stewed beef, and provides a tallowy, beeflike flavor character [91]. Although aldehydes pro-vide desirable flavor character to cooked meat, they can contribute rancid and "warmed-over" flavors at high concentrations, resulting from autoxida-tion of lipids [92].

Two fatty acids, 4-methyloctanoic and 4-methylnonanoic acid, provide the characteristic flavor of mutton [87]. 2-Pentylpyridine has been identified as the most abundant alkylpyridine isolated from roasted lamb fat. This compound has a fatty, tallowy aroma at an odor threshold of 0.6 ppb, and is suspected to negatively impact acceptance of lamb and mutton [87]. Rep-resentative structures for meat and seafood flavor impact compounds are shown in Figure 9.7.

The "fishy" aroma of seafood is incorrectly attributed to trimethyl amine. Flavor formation in fresh and saltwater fish results from complex enzymatic, oxidative, and microbial reactions of omega-3 polyunsaturated fatty acid

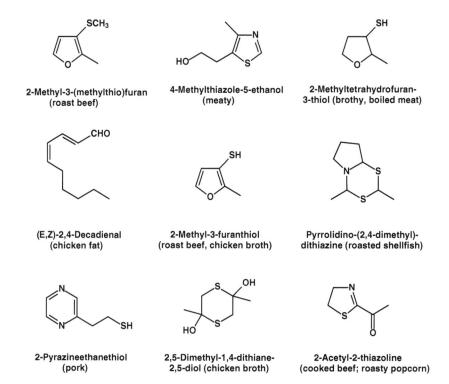

2-Methyl-3-(methylthio)furan
(roast beef)

4-Methylthiazole-5-ethanol
(meaty)

2-Methyltetrahydrofuran-
3-thiol (brothy, boiled meat)

(E,Z)-2,4-Decadienal
(chicken fat)

2-Methyl-3-furanthiol
(roast beef, chicken broth)

Pyrrolidino-(2,4-dimethyl)-
dithiazine (roasted shellfish)

2-Pyrazineethanethiol
(pork)

2,5-Dimethyl-1,4-dithiane-
2,5-diol (chicken broth)

2-Acetyl-2-thiazoline
(cooked beef; roasty popcorn)

Figure 9.7 Representative meat and seafood character-impact flavor compounds.

precursors. (e.g., eicosapentaenoic acid) [93–95]. Hence, fish flavor is mostly composed of noncharacterizing "planty" or "melonlike" aromas from lipid-derived unsaturated carbonyl compounds. Examples are (Z)-1,5-octa-dien-3-one (geranium-like) in boiled cod [96] and (E,Z)-2,6-nonadienal (cucumber-like) in boiled trout [97].

Several notable marine character-impact aroma compounds include 5,8,11-tetradecatrien-2-one, which exhibits a distinct seafood aroma character described as "cooked shrimplike" or "minnow bucket" [98]. A second example is an extremely potent odorant in cooked shellfish, including shrimp and clam, identified by Kubota and coworkers [99] as pyrroli-dino[1,2-e]-4H-2,4-dimethyl-1,3,5-dithiazine. This dithiazine contributes a roasted character to boiled shellfish and has the lowest odor threshold recorded to date, 10^5 ppt in water. 2,4,6-Tribromophenol and other bro-mophenol isomers have been associated with the ocean-, brine-, and iodine-like flavor character in seafood such as Australian ocean fish and prawns. The source of the bromophenols is thought to be polychaete worms, which form an important part of the diet for many fish and prawn species [100]. Dimethyl sulfide is reported to be the character aroma of stewed clams and oysters [93].

G. Cheese and dairy flavors

With a few exceptions, many of the known important flavors in dairy products do not provide characterizing roles. This is especially true for milk, cheddar cheese, and cultured products such as sour cream and yogurt. Delta-lactones are important flavors in butter, buttermilk, and cheeses, which are derived from triglycerides containing hydroxyl fatty acids. Although not directly contributing as dairy character-impact compounds, lactones play key supporting roles. The subject of key odor-active compounds in milk and dairy flavors has been recently reviewed [101–105]. A summary of charac-ter-impact compounds for cheese and dairy flavors is presented in Table 9.7.

Urbach [106] discusses the formation of volatile flavor compounds in different varieties of cheeses and provides a compilation of important aroma compounds. A recent qualitative assessment by Sable and Cottenceau [107] surveys the significant flavor volatiles that have been identified in soft mold-ripened cheeses including camembert, brie, blue, gorgonzola, munster, and limburger among others. 1-Octen-3-ol, 2-phenylethanol, and 2-phenyl-ethyl acetate are character-impact components in camembert-type cheese; these compounds together with sulfur compounds, 1-octen-3-one, and δ-decalactone are reported as the key aroma substances for camembert. 4-Methyl- and 4-ethyloctanoic acids contribute the "waxy or animal" char-acter in fresh chevre-style goat cheese flavor [108]. 2-Heptanone, 2-nonanone, and short- and moderate-chain fatty acids are the dominant character com-pounds of blue cheese flavor. Sulfur compounds, especially methanethiol, hydrogen sulfide, and dimethyl disulfide contribute to the strong garlic or putrid aroma of soft-smear or surface-ripened cheeses. Key aroma

Table 9.7 Character-Impact Flavor Compounds in Cheese and Dairy Products

Character-Impact Compounds	CAS Registry No.	Occurrence	Reference
2,3-Butanedione	[431-03-8]	Butter	116, 121
δ-Decalactone	[705-86-2]	Butter	116, 120, 121
6-Dodecen-γ-lactone	[156318-46-6]	Butter	116, 120
		Cheese, cheddar	112, 113
2-Heptanone	[110-43-0]	Cheese, Blue	106
1-Octen-3-ol	[3391-86-4]	Cheese, Camembert	107
Butryric acid	[107-92-6]	Cheese, cheddar	112
Methional	[3268-49-3]	Cheese, cheddar	112
		Parmigiano Reggiano	109-111
Skatole	[83-34-1]	Cheese, cheddar	112, 114
2-Acetyl-1-pyrroline	[85213-22-5]	Cheese, cheddar	112, 114
		Whey protein conc.	119
2-Acetyl-2-thiazoline	[22926-41-8]	Cheese, cheddar	112-114
Tetramethylpyrazine	[1124-11-4]	Cheese, cheddar	114
2,6-Dimethylpyrazine	[108-50-9]	Parmigiano Reggiano	110
		Whey, dried	118
Homofuraneol	[110516-60-4]	Cheese, Swiss	122
Propionic acid	[79-09-4]	Cheese, Swiss	122
		Cheese, Cheddar	112, 113
4-Methyloctanoic acid	[54947-74-9]	Cheese, goat	108
4-Ethyloctanoic acid	[16493-80-4]	Cheese, goat	108
(Z)-4-Heptenal	[6728-31-0]	Cream	4
		Cheese, cheddar	113
(E,E)-2,4-Nonadienal	[5910-87-2]	Cream	101
1-Nonen-3-one	[24415-26-7]	Yogurt; milk	123
Furaneol	[3658-77-3]	Milk, nonfat dry	117
		Whey protein conc.	118, 119
		Butter, heated	120, 121

compounds in Parmigiano Reggiano cheese were recently reported, including 3-/2-methylbutanal, 2-methylpropanal, dimethyltrisulfide, diacetyl, methional, ethyl C_4–C_8 fatty acid esters, and C_2–C_8 fatty acids [109–111]. In the basic fraction, 2,6-dimethylpyrazine and 6-ethyl-2,3,5-trimethylpyrazine had very strong "nutty," "baked" characters and the highest flavor intensities among the six pyrazines detected [110].

By a wide margin, cheddar is the most popular cheese flavor in North America. Although its flavor is described as "sweet, buttery, aromatic, and walnut," there is no general consensus among flavor chemists about the identity of individual compounds or groups of compounds responsible for cheddar flavor. At present, it is thought to arise from a unique balance of key volatile components, rather than a unique character-impact compound. Two recent sensory-guided flavor studies concluded that butyric acid, acetic acid, methional, homofuraneol (5-ethyl-4-hydroxy-2-methyl-(2*H*)-furan-3-one), (*E*)-2-nonenal, (*Z*)-4-heptenal, (*Z*)-1,5-octadien-3-one, and 2-acetyl-1-pyrroline are primary contributors to the pleasant mild flavor of cheddar cheese [112,113]. Important contributors to cheddar aroma are 2,3-butanedione, dimethyl sulfide, dimethyl trisulfide, and methanethiol [112]. A desirable "nutty" flavor supports "sulfur" and "brothy" characters in a quality aged cheddar cheese sensory profile. In these recent studies, 2-acetyl-2-thiazoline [112] and 2-acetyl-1-pyrroline [112,113] contribute "roasted, corny" flavors, which were suggested to be related to the "nutty" cheddar flavor. In an effort to characterize the "nutty" flavor in cheddar cheese, a comprehensive sensory analytical study concluded that Strecker aldehydes (2-methylpropanal and 2- and 3-methylbutanal), which individually are "green, malty, chocolate", provide an enhanced overall "nutty" flavor when added to cheddar cheese [114]. In the case of cheddar cheese powders, dimethyl sulfide imparts a desirable "creamed corn" flavor [115].

The most aroma-active compounds in fresh sour cream butter were elucidated as the character-impact compound diacetyl, with supporting roles from δ-decalactone, (*Z*)-6-dodeceno-γ-lactone, and butyric acid [116].

Key aroma-active compounds have been reported in dried dairy products including nonfat milk and whey powders. Furaneol is the primary contributor to the "sweet, caramelized" aroma character of nonfat dry milk, with supporting roles from heat-generated sotolon, maltol, vanillin, and (*E*)-4,5-epoxy-(*E*)-2-decenal [117]. Compounds important to dried sweet whey from cheddar cheese manufacture include butyric and hexanoic acids, maltol, Furaneol, dimethylpyrazines, 2,3,5-trimethylpyrazine, and several lactones [118]. Additional flavor-impact compounds identified in whey protein concentrate and whey protein isolate were butyric acid (cheesy), 2-acetyl-1-pyrroline (popcorn), 2-methyl-3-furanthiol (brothy or burnt), Furaneol (maple/spicy), 2-nonenal (fatty), (*E,Z*)-2,6-nonadienal (cucumber), and (*E,Z*)-2,4-decadienal (fatty or oxidized) [119].

Representative structures of significant or newly identified flavorants including homofuraneol in mild cheddar cheese [112, 113] and Swiss cheese [122], δ-decalactone in butter and buttermilk [116, 120, 121], (*Z*)-6-dodecen-γ-lactone in butter [116,120] and mild cheddar cheese [112,113], and 1-nonen-3-one in milk and yogurt [123] are shown in Figure 9.8.

III.   Characterizing aromas in off-flavors

Flavor defects, so-called taints, malodors, or "off-flavors," are sensory attributes that are not associated with the typical aroma and taste of foods and beverages. These defects can range from subtle to highly apparent, and are often significant detractors to food quality. Off-flavors can be produced due to several possible factors: contamination (air, water, packaging, or shipping materials), ingredient mistakes in processing or generation (chemical or microbial) in the food itself. In the latter instance, generation of off-flavors

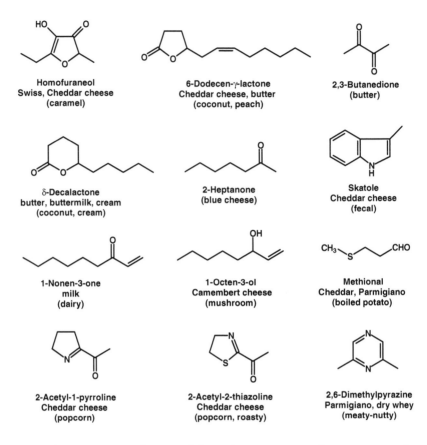

Figure 9.8 Representative cheese and dairy character-impact flavor compounds.

in foods may result from oxidation, nonenzymatic browning, chemical reactions between food constituents, light-induced reactions, or enzymatic pathways.

Over the past 15 years, numerous complete and detailed reviews have discussed the occurrence of off-flavors in food and packaging systems [124–128]. The intent of this section is to summarize recent highlights and off-flavorants of significance, without striving for comprehensiveness.

Lipid-derived volatile compounds play an important role in many food flavors. These compounds contribute to the characteristic and desired flavor attributes of foods, but can also cause off-flavors depending on their concentrations relative to other sensorially-relevant odorants. For example, the "cardboard" off-flavor of butter oil is primarily related to (*E*)-2-nonenal, which is formed by the autoxidation of palmitoleic acid [129]. Carbonyl compounds formed by lipid peroxidation were identified in cooked beef, which develops a warmed-over flavor from reheating after 2-day refrigerated storage [92]. Warmed-over flavor is principally caused by formation of (*E*)-4,5-epoxy-(*E*)-2-decenal (metallic) and hexanal (green) notes, which are not present in freshly cooked beef. Similarly for boiled chicken, "green-card-board-like metallic" off-odors were formed during refrigerated storage and reheating, primarily from a sevenfold increase in hexanal [90]. These and other unsaturated carbonyl compounds, including (E,E)-2,4-decadienal (deep-fried) and 1-octen-3-one (mushroom-like/beany), are primarily responsible for rancid off-flavors in soybean and canola oils through oxidation of linoleic and linolenic acids [130]. The light-induced 3-methyl-2,4-nonanedi-one (strawy, lardlike, beany) strongly contributes to the reverted off-flavor of oxidized soybean oil, because of its extremely low odor threshold (0.007 to 0.014 ppt) [131]. Melon odors are associated with foods cooked in partially hydrogenated soybean oils that have undergone oxidative deterioration during heating. 6-Nonenal (cucumber or melon) from (Z,Z)-9,15-linoleic acid is reported to be a character flavor associated with these deteriorated oils [132]. (Z)-3-Hexenal and (Z,Z)-3,6-nonadienal were shown to contribute substantially to the fatty, fishy off-flavor of boiled trout, which was in frozen storage for several months before cooking [97]. Other carbonyl compounds that are likely to contribute to characteristic "fishy" off-flavors in oxidized seafood because of their low odor thresholds include 2,6-nonadienal and 1,5-octa-dien-3-one [93]. A summary of off-flavor-impact compounds for foods, beverages, and packaging materials is presented in Table 9.8, with representative chemical structures in Figure 9.9.

trans-2-Nonenal is considered to be the characteristic volatile responsible for a "stale, cardboard" off-flavor in aged packaged beer. Mechanistic studies using labeled nonenal confirmed that cardboard off-flavor in finished beer arises from lipid autoxidation during wort boiling and not from lipoxygenase activity during the mashing step [133]. Studies also revealed that 2-furfuryl ethyl ether is responsible for an astringent, stale off-flavor in beer [134]. Stale flavor was reproduced by adding the furfuryl ether and trans-2-nonenal to fresh beer but not when either compound was added individually. Lipid-derived (E,E)-2,4-decadienal, in addition to hexanal, (E)-2-octenal, and (E)-2-nonenal, were shown to be the most potent off-flavor compounds in precooked vacuum-packaged potatoes [135]. In dry, raw spinach, (Z)-1,5-octa-dien-3-one and methional, are responsible for a "fishy" off-flavor at a 1:100 ratio. 3-Methyl-2,4-nonanedione produces a "haylike" off-flavor character from the oxidative degradation of furanoid fatty acids in dry spinach [136] and in dry parsley [137].

Exposure of beer to light has been shown to produce 3-methyl-2-butene-1-thiol, which produces a skunky off-flavor in "sun-struck" or "light-struck" ales [138,139]. This mercaptan has a sensory threshold of 0.05 ppb in beer. It results from complex photo-induced degradations of isohumulones (hop-derived bitter iso-acids) to form free-radical intermediates, which subsequently react with the thiol group of cysteine. Light-struck off-flavor can be controlled in beer through packaging technology (colored glass bottles), use of chemically-modified hop-derived bitter acids, antioxidants, or its precipitation with high-molecular-weight gallotannins by addition of zinc salts [140]. In addition to dimethyl sulfide, thioesters have been reported to

Table 9.8 Off-Flavor-Impact Compounds in Food and Beverage Products

Impact Compounds	CAS Registry No.	Off-Flavor	Occurrence	Reference
Geosmin	[19700-21-1]	Musty, earthy	Catfish, wheat, water	150,154
2,4,6-Trichloroanisole	[87-40-1]	Musty, moldy	Coffee, wine corks	153,154
2-Methylisoborneol	[2371-42-8]	Earthy, musty	Coffee, catfish, beans	153,150,154
2-iso-Propyl-3-methoxypyrazine	[25773-40-4]	Peasy	Coffee, cocoa beans	153,126
4,4,6-Trimethyl-1,3-dioxane	[1123-07-5]	Musty	Packaging film	157
8-Nonenal	[39770-04-2]	"Smoky" plastic	Polyethylene packaging	158
2-Iodo-4-methylphenol	[16188-57-1]	Medicinal	Lemon cake mix	156
(E)-2-Nonenal	[2463-53-8]	Cardboard, stale	Beer, packaged	138
		Cardboard	Oxidized butter	129
		Cardboard	Soybean lecithin	130
(E)-4,5-Epoxy-(E)-2-decenal	[134454-31-2]	Metallic	Warmed-over (beef)	92
		Metallic	Oxidized soybean oil	130
		Metallic	Butter	60
(E,Z)-2,6-nonadienol	[7786-44-9]	Metallic	Buttermilk	168
1-Octen-3-one	[4312-99-6]	Metallic, mushroom	Butterfat	162
		Beany	Rancid soybean oil	130
Hexanal	[66-25-1]	Green grass	Rancid soybean oil	130
		Rancid	Warmed-over (beef)	92
(E,E)-2,4-Decadienal	[25152-84-5]	Deep-fried	Rancid soybean oil	130
2,3-Diethyl-5-methylpyrazine	[18138-04-0]	Roasty, earthy	Soybean lecithin	130
6-Nonenal	[6728-35-4]	Cucumber, melon	Part-hydrogenated soy oil	132
(Z)-3-Hexenal	[6789-80-6]	Fatty, fishy	Aged trout	97
(Z,Z)-3,6-Nonadienal	[21944-83-2]	Fatty, fishy	Aged trout	97
5α-Androst-16-en-3-one	[18339-16-7]	Urine	Boar meat	126,177
Skatole	[83-34-1]	Fecal-like	Boar meat; potato chips	177, 126
		Medicinal	Beef	178

Compound	CAS	Flavor	Source	Reference
2-Furfuryl ethyl ether	[6270-56-0]	Stale, astringent	Beer	134
3-Methyl-2-butene-1-thiol	[5287-45-6]	Skunky	Beer (light-struck)	138,139
S-Methyl hexanethioate	[2432-77-1]	Cabbagy, rubbery	Beer	19a
4-Mercapto-4-methyl-2-pentanone	[19872-52-7]	Cat urine, ribes	Beer	141,142
Methional	[3268-49-3]	Worty	Beer (alcohol-free)	144
		Cooked vegetables	Oxidized white wine	143
		Sunlight off-flavor	Milk	127,164
		Potato	UHT-milk	173
(E)-1,3-Pentadiene	[504-60-9]	Kerosene	Cheese (sorbic acid)	126,169
2,4,5-Trimethyloxazole	[20662-84-4]	Melon, kiwi	Dried sour cream	115
Tetradecanal	[124-25-4]	Sickening, aldehydic	Milk powder	170
β-Ionone	[79-77-6]	Hay-like	Milk powder	170
Benzothiazole	[95-16-9]	Sulfuric, quinoline	Milk powder	170
2,6-Dimethylpyrazine	[108-50-9]	Cooked milk	UHT-milk	173
2-Ethyl-3-methylpyrazine	[15707-23-0]	Cooked milk	UHT-milk	173
2-Aminoaceto-phenone	[551-93-9]	Gluey, glutinous	Milk powder, casein	159, 160, 171
		UTA	White wine	161
3-Methyl-2,4-nonanedione	[113486-29-6]	Strawy, beany	Soybean oil (light-induced)	131
		Hay-like	Dried spinach, parsley	136,137
(Z)-1,5-Octadien-3-one	[65767-22-8]	Fishy	Dried spinach, old fish	136,93
Bis(2-methyl-3-furyl)disulfide	[28588-75-2]	Vitamin B_1 odor	Thiamin degra-dation	166
4-Methyl-2-isopropylthiazole	[15679-13-7]	Vitamin, cabbage	Orange juice (Vit. B_2)	146
Sotolon	[28664-35-9]	Burnt, spicy	Citrus soft drink	165
(S)-(+)-Carvone	[2244-16-8]	Woody, terpeny	Orange juice (oxidized)	179
4-Vinylguaiacol	[7786-61-0]	Rotten, old fruit	Orange juice; apple juice	180, 181
			Beer, wort	18

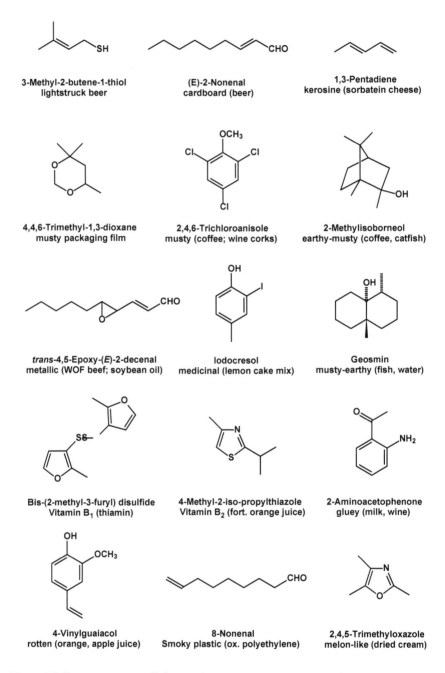

Figure 9.9 Representative off-flavor character-impact compounds.

contribute a cabbagy, rubbery off-note that sometimes occurs in beer, the most significant being S-methyl hexanethioate, which has a detection threshold of 1 ppb [19a]. Diacetyl can produce an undesirable buttery off-character in beer through accelerated fermentation, whereby brewer's yeast has not converted all of the diketone intermediates to flavor-inactive acetoin and 2,3-butanediol [138]. Another staling off-flavor in beer is described as "catty or ribes", whose character comes from 4-mercapto-4-methyl-2-pentanone (cat ketone). This off-flavor can develop rapidly in beers that are packaged and stored with a high air content in the headspace [141]. Cat ketone was also reported in a "catty" off-flavored beer prepared in a freshly painted malting facility, in which mesityl oxide was an impurity in the paint [142]. Mesityl oxide is an intermediate in the thioketone formation (see Figure 9.3).

Strecker aldehydes are a frequent source of off-flavors in fermented products. Development of off-flavors in oxidized white wines typically marks the end of shelf life. Methional (3-methylthiopropionaldehyde) was identified as producing a "cooked vegetables" off-flavor character in a young white wine that had undergone spontaneous oxidation [143]. Methional levels increased in wines spiked with methionol or methionine, suggesting its formation via direct peroxidation or Strecker degradation of methionine. Methional was recently demonstrated to impart a "worty" off-flavor in alcohol-free beer, with more sensory significance than was previously attributed to 3-methyl- and 2-methylbutanal for this off-taste [144]. In an alcohol-free beer medium, perception of worty off-flavors is strengthened by the absence of ethanol and by the higher level of sugars. Finally, methional is responsible for "sunlight" flavor in photooxidized milk, resulting from the oxidation of methionine in the presence of riboflavin as a sensitizer [145]. Another vitamin-derived off-odor problem in which a pineapple fruit juice beverage was fortified with riboflavin was recently described. The "vitamin, cabbage, brothy, vegetable soup" off-odor was characterized as 4-methyl-2-isopropylthiazole, which resulted from riboflavin-sensitized Strecker degradation of valine, cysteine, and methionine, followed by reaction of the resulting aldehydes with ammonia and hydrogen sulfide [146].

Sulfur compounds present in wine can have a detrimental effect on aroma character, producing odors described as "garlic, onion, and cauliflower," the so-called Boeckser aromas. This sulfurous character is correlated with with 2-methyl-3-hydroxythiophene, 2-methyl-3-furanthiol, and ethanethiol, and their concentrations in wine are influenced by winery procedures and the use of certain winemaking yeasts [147]. Off-flavors in European wines were associated with the nonvolatile *bis*(2-hydroxyethyl) disulfide, a precursor to the "poultry-like" character of 2-mercaptoethanol and hydrogen sulfide [148].

Other undesirable "Brett" off-flavors are produced in wines owing to contamination in grape musts and wineries with *Brettanomyces* yeast strains, named after their original discovery in British beers. These off-flavors are typically are described as "Band-aid," "barnyard," "horse sweat," "plastic," and "wet animal." A recent study identified key character compounds in high-Brett wines as isovaleric acid, 4-ethyl guaiacol, and 4-ethyl phenol; a potent "plastic" note was uncharacterized [149].

Musty aromas and flavors are a major problem in a variety of foods and packaging materials, and many of the causative character-impact materials have powerful impact [150]. For example, 2,4,6-trichoroanisole is a highly odorous metabolite of a fungus that attacks wood, paperboard packaging, and wine corks [125,126]. Trichloroanisole has a musty, haylike odor, and possesses an extremely low odor threshold (~0.05 ppt). Both choloroanisoles and their corresponding phenols can provide musty off-flavor character to foods and process water. For example, chloroanisoles were related to the musty off-flavor of raisins [151]. About 20% of Brazilian coffee production exhibits the so-called "Rioy defect," characterized by a strong off-flavor that is often described as "medicinal, phenolic, or iodine-like" [152]. Occasionally, this defect also occurs in coffees from other origins. 2,4,6-Trichloroanisole in concentrations ranging from 1 to 100 ppb was identified as the most likely key compound for the Rioy off-flavor as analyzed by capillary GC, GC-sniffing, and GC–MS. 2,4,6-Trichlorophenol, the probable precursor, was also found in most of these samples. Adding trichloroanisole to freshly brewed coffee imparted to it the same off-flavor notes as described in actual Rioy coffee. Its perception threshold in coffee brew was found to be 8 ppt by aroma and 1 to 2 ppt for taste [152]. In Robusta coffees, 2-methylisoborneol provides an "earthy, tarry" character at 5 ppt, and must be removed during processing to approach the flavor of Arabica coffees [153]. Geosmin (1,10-dimethyl-9-decalol) imparts an earthy–musty off-odor to drinking water, and a "muddy" odor to catfish and tilapia fish, which live in brackish water [150,154,155]. It is produced from *actinomycetes* microorganisms in soil, planktonic algae, and fungi. 2-Methylisoborneol also provides an earthy-musty character to catfish [154]. A "medicinal" off flavor was produced in a lemon-flavored cake mix by reaction between two minor ingredients, *p*-cresol from the lemon flavor and iodine from iodized salt. The resulting iodocresol (2-iodo-4-methylphenol) was shown to possess a "medicinal" character at 0.2 ppb aroma threshold [156]. McGorrin et al. studied a "musty" off-odor in printed plastic film which was not a phenol as expected, but was identified as 4,4,6-trimethyl-1,3-dioxane [157]. It was formed during film manufacture as a reaction product between 2-methyl-2,4-pentanediol, a solvent coating to facilitate ink adhesion, and formaldehyde, a component in the ink.

Another common source of packaging-derived off-flavors is from polyethylene packaging materials, which can sometimes transfer undesirable off-flavors to foods stored in them. The so-called "plastic" or "smoky-poly" off-flavors can be generated by minor plasticizer components in the resin, or from oxidation of the polyolefin during polymerization and subsequent thermal extrusion into films, sheets, or containers. Polyethylene films, which are manufactured under high-temperature conditions, are more likely to contribute these off-flavors. Recently, a previously unidentified potent "plastic" off-flavorant was identified as 8-nonenal, which contributes a characteristic "plastic" off-flavor to snack foods packaged in oxidized polyethylene containers [158].

The same compounds that have positive character impact can become off-flavors in different food contexts. For example, indole and skatole in cheddar cheese flavor become "fecal" in the context of potato chips [126]. 2-Aminoacetophenone ("foxy" character) in Concord grape, imparts a "gluey" flavor in milk powder and casein through degradation of tryptophan [159,160]. In white wines (such as Riesling), it contributes an "untypical ageing note" from either tryptophan or kynurenine precursors [161]. 1-Octen-3-one (mushroom) imparts an undesired "metallic" flavor in dairy products and oxidized vegetable oils [162]. A recent study identified two key "metallic" odorants generated in aqueous ferrous sulfate solutions, in which 1-nonen-3-one provides a 10-fold more potent "metallic-smelling" character relative to 1-octene-3-one [163]. Sunlight off-flavor in milk (cardboard-like) can result from photooxidation of milk exposed to high-intensity fluorescent light or sunlight, which converts methionine to methional, a flavor-impact compound in boiled potatoes. Methional is further degraded to other impact sulfur compounds including dimethyl disulfide [164]. Sotolon, the furanone character-impact flavor of fenugreek (Table 9.1) has recently been shown to cause a "burnt, spicy" off-flavor in citrus soft drinks, generated by reaction of ascorbic acid with ethanol [165]. Whereas 2-isopropyl-3-methoxy pyrazine contributes to pea and earthy potato flavor character, it provides an undesirable "peasy" off-flavor to Rwandan coffee [153] and fermented cocoa beans [126]. Although *bis*(2-methyl-3-furyl)disulfide contributes a desirable aged, prime rib flavor in beef, it is the principal "vitamin B" off odor resulting from thiamin degradation [166].

Character compounds which contribute positive flavor impact at low levels can become off-flavors when they occur at higher concentrations. For example, dimethyl sulfide provides an appropriate "cornlike" background character to beer flavor at low levels; however, it contributes a highly undesirable "cooked vegetable" or "cabbage-like" malodor when present at levels significantly above its sensory threshold (30 to 45 ppb) [167]. Similarly for cheddar cheese flavor, it imparts a rotten vegetable taste when present at high levels [113].

Although off-flavors in dairy products have been reviewed [126,127], there are several recent developments. In sour-cream buttermilk, the key odorant responsible for a metallic off-flavor was identified as (*E,Z*)-2,6-nonadienol [168]. During cream fermentation, its formation occurs from peroxidation of alpha-linolenic acid to generate the 2,6-nonadienal precursor, with subsequent reduction to the dienol by starter culture reductases that remain active during storage. Metallic off-flavors are not formed readily in fermented sweet-cream buttermilk because of significantly lower concentrations of alpha-linolenic acid [168]. The common use of sorbic acid or potassium sorbate as a mold inhibitor in commercial dairy products often produces an off-flavor described as "kerosene, plastic, or paintlike", which may incorrectly be attributed to packaging materials. The source of the taint is (*E*)-1,3-pentadiene, which results from decarboxylation of sorbates by lactic acid bacteria in yogurt, cheese, and margarine [126,169]. A "melon, ripe kiwifruit" off-flavor in spray-dried cultured dairy products was identified as 2,4,5-trimethyloxazole, a product of the Maillard reaction between diacetyl and arginine (whey source), or acetaldehyde with ammonia [115].

Other significant recurring themes in dairy taints include studies on the characteristic off-flavor in spray-dried skim milk powder, which was related to contributions from tetradecanal (sickening, aldehydic), β-ionone (hay-like), and benzothiazole (sulfuric, quinoline) at low parts per billion levels [170]. An intense "musty, stale" off-flavor was identified as 2-aminoacetophenone in micromilled milk powder [171]. A recent AEDA study of the characteristic aroma components of dried rennet casein indicated that, while 2-aminoacetophenone is a potent odorant, sensory reconstitution confirmed that hexanoic acid, indole, guaiacol, and *p*-cresol were the principal factors for its typical "animal or wet dog" odor [172]. In ultrahigh-temperature (UHT) processed milk, the "UHT milk flavor" character is contributed by 2,6-dimethylpyrazine, 2-ethylpyrazine, 2-ethyl-3-methylpyrazine, and methional [173]. Additionally, dimethyl sulfide, diacetyl, 2-heptanone, 2-nonanone, 2-methylpropanal, 3-methylbutanal, nonanal, and decanal were reported as significant flavor components of UHT milk off-flavor [174]. Recently, formation of the UHT off-flavor was shown to be inhibited by addition of 0.1% epicatechin during milk processing [175]. Fruity (pineapple-like) off-flavors in pasteurized milk indicate the presence of high levels of ethyl butyrate, ethyl hexanoate, ethyl octanoate, and ethyl decanoate esters, whereas rancid, soapy tastes arise from decanoic and dodecanoic acids [176].

As previously discussed, lipid oxidation is generally related to flavor deterioration in meat and meat products. However, "boar taint", an intense urine-like off-odor, is attributed primarily to the flavor synergy between two compounds in boar fat, androstenone (5-α-androst-16-ene-3-one) from testes, and the "fecal" skatole (3-methyl indole) from tryptophan breakdown [177]. Generally, women tend to be more sensitive to the odor than men. Skatole has also been implicated in a "medicinal" off-odor in beef [178].

Off-flavors in citrus oils such as orange oil have been related to autoxidation of limonene to (S)-carvone (caraway; Figure 9.1) and carveol, producing "woody, turpeny" off-flavors [179]. Other off-flavors in orange juice arise from 4-vinylguaiacol, which contributes an "old fruit" and "rotten" character because of degradation of ferulic acid [180]. Conversion of ferulic acid to 4-vinylguaiacol by yeast contaminants, with corresponding off-flavor development, has been reported in unpasteurized apple juice [181] and in beers and worts [182]. Thermal abuse during processing or elevated temperature storage of orange and grapefruit juice produces Furaneol, which contributes a "sweet, pineapple" defect [180].

IV. Conclusion

The objective of this chapter was to provide an updated review of character-impact compounds in flavors and off-flavors. Particular emphasis was placed on compounds that have been identified in natural flavor systems. The summarized data can be applied in creative flavor compounding efforts to replicate and monitor production of flavors and in food processing to ensure flavor quality.

References

1. L.M. Nijssen, C.A. Visscher, H. Maarse, L.C. Willemsens, and M.H. Boelens, Eds., *Volatile Compounds in Food. Qualitative and Quantitative Data*, 7th ed., TNO Biotechnology and Chemistry Institute, Zeist, The Netherlands, 1996.
2. S.S. Chang, Food flavors, *Food Technol.* 43(12): 99, 1989.
3. R. Emberger, An analytical approach to flavor research, *Cereal Foods World*, 30(10): 691, 1985.
4. F. Fischetti, Flavors and spices, in *Kirk-Othmer Encyclopedia of Chemical Technology*, 4th ed., Vol. 11, M. Howe-Grant, Ed., 1994, p. 16.
5. P. Darriet, T. Tominaga, V. Lavigne, J.-N. Biodron, and D. Dubourdieu, Identification of a powerful aromatic component of *Vitis vinifera* L. var. Sauvignon wines: 4-mercapto-4-methylpentan-2-one, *Flav. Frag. J.* 10: 385, 1995.
6. R.J. McGorrin, Character impact compounds: flavors and off-flavors in foods, in *Flavor, Fragrance, and Odor Analysis*, R. Marsili, Ed., Marcel Dekker, New York, 2002, pp. 375–413.
7. H. Maarse, Introduction, in *Volatile Compounds in Foods and Beverages*, H. Maarse, Ed., Marcel Dekker, New York, 1991, p. 1.
8. W. Grosch, Flavor of coffee: a review, *Nahrung*, 42: 344, 1998.
9. A.R. Mayol and T.E. Acree, Advances in gas chromatography-olfactometry, in: *Gas Chromatography-Olfactometry: The State of the Art*, J.V. Leland, P. Schieberle, A. Buettner, and T.E. Acree, Eds., ACS Symposium Series 782, American Chemical Society, Washington, DC, 2001, p. 1.
10. B.H. Mistry, T. Reineccius, and L.K. Olson, Gas chromatography-olfactometry for the determination of key odorants in foods, in *Techniques for Analyzing Food Aroma*, R. Marsili, Ed., Marcel Dekker, New York, 1997, p. 265.
11. W. Pickenhagen, Flavor chemistry — the last 30 years, in *Flavor Chemistry: Thirty Years of Progress*, R. Teranishi, E. L. Wick, and I. Hornstein, Eds., Kluwer Academic/Plenum, New York, 1999, p. 75.
12. M.H. Boelens, Spices and condiments II, in *Volatile Compounds in Foods and Beverages*, H. Maarse, Ed., Marcel Dekker, New York, 1991, p. 449.
13. M. Diaz-Maroto, I. Hidalgo, E. Sanchez-Palomo, and M. Perez-Coello, Volatile components and key odorants of fennel (*Foeniculum vulgare* Mill.) and thyme (*Thymus vulgaris* L.) oil extracts obtained by simultaneous distillation-extraction and supercritical fluid extraction, *J. Agric. Food Chem.* 53: 5385, 2005.
14. K. Bauer, D. Garbe, and H. *Surburg, Common Fragrance and Flavor Materials*, 3rd ed., Wiley-VCH, Weinheim, Germany, 1997.
15. I. Blank, J. Lin, S. Devaud, R. Fumeaux, and L. B. Fay, The principal flavor components of fenugreek (*Trigonella foenum-graecum* L.), in *Spices: Flavor Chemistry and Antioxidant Properties*, S.J. Risch and C.-T. Ho, Eds., ACS Symposium Series 660, American Chemical Society, Washington, D.C., 1997, p. 12.
16. R.D. Hiserodt, C.-T. Ho, and R.T. Rosen, The characterization of volatile and semivolatile components in powdered turmeric by direct thermal extraction gas chromatography-mass spectrometry, in *Spices: Flavor Chemistry and Antioxidant Properties*, S.J. Risch and C.-T. Ho, Eds., ACS Symposium Series 660, American Chemical Society, Washington, D.C., 1997, p. 80.
17. K.R. Cadwallader, H.H. Baek, and M. Cai, Characterization of saffron flavor by aroma extract dilution analysis, in *Spices: Flavor Chemistry and Antioxidant Properties*, S.J. Risch and C.-T. Ho, Eds., ACS Symposium Series 660, American Chemical Society, Washington, D.C., 1997, p. 66.

18. D. Rowe, More Fizz for your buck: high-impact aroma chemicals, *Perf. Flav.* 25: (9/10), 1, 2000.
19. (a). G. Ohloff, I. Flament, and W. Pickenhagen, Flavor chemistry, *Food Rev. Int.* 1: 99, 1985. (b). T. Jagella and W. Grosch, Flavor and off-flavor compounds of black and white pepper (*Piper nigrum* L.). II. Odor activity values of desirable and undesirable odorants of black pepper, *Eur. Food Res. Technol.* 209: 22, 1999.
20. I. Blank, A. Sen, and W. Grosch, Sensory study on the character-impact flavor compounds of dill herb (*Anethum graveolens* L.), *Food Chem.*, 43: 337, 1992.
21. M.H. Boelens and L.J. van Gemert, Volatile character-impact sulfur compounds and their sensory properties, *Perf. Flav.*, 18(3): 29, 1993.
22. W.M. Randle, Onion flavor chemistry and factors influencing flavor intensity, in *Spices: Flavor Chemistry and Antioxidant Properties*, S.J. Risch and C.-T. Ho, Eds., ACS Symposium Series 660, American Chemical Society, Washington, D.C., 1997, p. 41.
23. E. Block, S. Naganathan, D. Putman, and S.-H. Zhao, Allium chemistry: HPLC analysis of thiosulfinates from onion, garlic, wild garlic (Ramsoms), leek, scallion, shallot, elephant (great-headed) garlic, chive, and Chinese chive, *J. Agric Food Chem.*, 40: 2418, 1992.
24. S. Widder, C.S. Luntzel, T. Dittner, and W. Pickenhagen, 3-Mercapto-2-methyl-pentan-1-ol, a new powerful aroma compound, *J. Agric. Food Chem.* 48: 418, 2000.
25. M.A. Majcher and H.J. Jelen, Identification of potent odorants formed during the preparation of extruded potato snacks, *J. Agric. Food Chem.* 53: 6432, 2005.
26. R.C. Berger, Fruits I, in *Volatile Compounds in Foods and Beverages*, H. Maarse, Ed., Marcel Dekker, New York, 1991, p. 283.
27. D.D. Roberts and T.E. Acree, Developments in the isolation and characterization of β-damascenone precursors from apples, in *Fruit Flavors: Biogenesis, Characterization, and Authentication*, R.L. Rouseff and M.M. Leahy, Eds., ACS Symposium Series 596, American Chemical Society, Washington, D.C., 1995, p. 190.
28. M. Larsen, L. Poll, and C.E. Olsen, Evaluation of the aroma composition of some strawberry (*Fragaria ananassa* Duch) cultivars by use of odor threshold values, *Z. Lebensm. Unters. Forsc.* 195: 536, 1992.
29. H.H. Baek and K.R. Cadwallader, Contribution of free and glycosidically bound volatile compounds to the aroma of muscadine grape juice, *J. Food Sci.* 64: 441, 1999.
30. P. Schieberle and T. Hofmann, Evaluation of the character impact odorants in fresh strawberry juice by quantitative measurements and sensory studies on model mixtures, *J. Agric. Food Chem.* 45: 227, 1997.
31. M.G. Kolor, Identification of an important new flavor compound in Concord grapes, *J. Agric. Food Chem.* 31: 1125, 1983.
32. A.L. Waterhouse and S.E. Ebeler, Eds., *Chemistry of Wine Flavor*, ACS Symposium Series 714, American Chemical Society, Washington, D.C., 1998.
33. Y. Fang and M. Qian, Aroma compounds in Oregon Pinot Noir wine determined by aroma extract dilution analysis (AEDA), *Flav. Frag. J.* 20: 22, 2005.
34. E. Campo, V. Ferreira, A. Escudero, and J. Cacho, Prediction of the wine sensory properties related to grape variety from dynamic-headspace gas chromatography-olfactometry data, *J. Agric. Food Chem.* 53: 5682, 2005.
35. L. Farina, E. Boido, F. Carrau, G. Versini, and E. Dellacassa, Terpene compounds as possible precursors of 1,8-cineole in red grapes and wines, *J. Agric. Food Chem.* 53: 1633, 2005.

36. (a) H. Guth, Identification of character impact odorants of different white wine varieties, *J. Agric. Food Chem.* 45: 3022, 1997. (b) H. Guth, Quantitation and sensory studies of character impact odorants of different white wine varieties, *J. Agric. Food Chem.* 45: 3027, 1997.

37. K. Kumazawa, K. Kubota, and H. Masuda, Influence of manufacturing conditions and crop season on the formation of 4-mercapto-4-methyl-2-pentanone in Japanese green tea (sen-cha), *J. Agric. Food Chem.* 53: 5390, 2005.

38. A. Buettner and P. Schieberle, Evaluation of key compounds in hand-squeezed grapefruit juice (*Citrus paradisi* Macfayden) by quantitation and flavor reconstitution experiments, *J. Agric. Food Chem.* 49: 1358, 2001.

39. R. Tressl, D. Bahri, and M. Kossa, Formation of off-flavor components in beer, in *The Analysis and Control of Less Desirable Flavors in Foods and Beverages*, G. Charlambous, Ed., Academic Press, New York, 1980, p. 293.

40. R.K. Boccorh, A. Paterson, and J.R. Piggot, Extraction of aroma components to quantify overall sensory character in a processed blackcurrant (*Ribes nigrum* L.) concentrate, *Flav. Frag. J.* 17: 385, 2002.

41. E. Demole, P. Enggist, and G. Ohloff, 1-p-Menthene-8-thiol: a powerful flavor impact constituent of grapefruit juice (*Citrus paradisi* MacFayden), *Helv. Chim. Acta* 65: 1785, 1982.

42. K. Mahattanatawee, R. Rouseff, M.F. Valim, and M. Naim, Identification and aroma impact of norisoprenoids in orange juice, *J. Agric. Food Chem.* 53: 393, 2005.

43. P.E. Shaw, Fruits II, in *Volatile Compounds in Foods and Beverages*, H. Maarse, Ed., Marcel Dekker, New York, 1991, p. 305.

44. H.-S. Choi, Characteristic odor components of kumquat (*Fortunella japonica* Swingle) peel oil, *J. Agric. Food Chem.* 53: 1642, 2005.

45. M. Larsen and L. Poll, Odor thresholds of some important aroma compounds in raspberries, *Z. Lebensm. Unters. Forsch.* 191:129, 1990.

46. K. Klesk, M. Qian, and R.R. Martin, Aroma extract dilution analysis of cv. Meeker (*Rubus idaeus* L.) red raspberries from Oregon and Washington, *J. Agric. Food Chem.* 52: 5155, 2004.

47. K. Klesk and M. Qian, Aroma extract dilution analysis of cv. Marion (*Rubus* spp. hyb) and cv. Evergreen (*R. laciniatus* L.) blackberries, *J. Agric. Food Chem.* 51: 3436, 2003.

48. Y. Wang. C. Finn, and M. Qian, Impact of growing environment on Chickasaw blackberry (*Rubus* L.) aroma evaluated by gas chromatography olfactometry dilution analysis, *J. Agric. Food Chem.* 53: 3563, 2005.

49. R.G. Buttery, Quantitative and sensory aspects of flavor of tomato and other vegetables and fruits, in *Flavor Science: Sensible Principles and Techniques*, T.E. Acree and R. Teranishi, Eds., ACS Books, American Chemical Society, Washington, D.C., 1993, p. 259.

50. P. Schieberle, S. Ofner, and W. Grosch, Evaluation of potent odorants in cucumbers (*Cucumis sativus*) and muskmelons (*Cucumis melo*) by aroma extract dilution analysis, *J. Food Sci.* 55: 193, 1990.

51. M. Guntert, G. Krammer, H. Sommer, and P. Werkhoff, The importance of the vacuum headspace method for the analysis of fruit flavors, in *Flavor Analysis: Developments in Isolation and Characterization*, C.J. Mussinan and M.J. Morello, Eds., ACS Symposium Series 705, American Chemical Society, Washington, D.C., 1998, p. 38.

52. F.B. Whitfield and J.H. Last, Vegetables in *Volatile Compounds in Foods and Beverages*, H. Maarse, Ed., Marcel Dekker, New York, 1991, p. 203.
53. G. Takeoka, Flavor chemistry of vegetables, in *Flavor Chemistry: Thirty Years of Progress*, R. Teranishi, E.L. Wick, and I. Hornstein, Eds., Kluwer Academic/Plenum, New York, 1999, p.287.
54. R.J. McGorrin and L. Gimelfarb, Comparison of flavor components in fresh and cooked tomatillo with red plum tomato, in *Food Flavors: Formation, Analysis, and Packaging Influences*, E.T. Contis, C.-T. Ho, C.J. Mussinan, T.H. Parliment, F. Shahidi, and A.M. Spanier, Eds., Elsevier, New York, Developments in Food Science Vol. 40, 1998, p. 295.
55. H.S. Marks, J.A. Hilson, H.C. Leichtweis, and G.S. Stoewsand, S-Methylcysteine sulfoxide in *Brassica* vegetables and formation of methyl methanethiosulfinate from Brussels sprouts, *J. Agric. Food Chem.* 40: 2098, 1992.
56. H.-W. Chin and R.C. Lindsay, Mechanisms of formation of volatile sulfur compounds following the action of cysteine sulfoxide lyases, *J. Agric. Food Chem.* 42: 1529, 1994.
57. R.K. Wagner and W. Grosch, Key odorants of French fries, *J. Am. Oil Chem. Soc.* 75: 1385, 1998.
58. C. Masanetz and W. Grosch, Key odorants of parsley leaves (*Petroselinum crispum* [Mill.] Nym. ssp. crispum) by odor-activity values, *Flav. Frag. J.* 13: 115, 1998.
59. R. Scarpellino and R.J. Soukup, Key flavors from heat reactions of food ingredients in *Flavor Science: Sensible Principles and Techniques*, T.E. Acree and R. Teranishi, Eds., ACS Books, American Chemical Society, Washington, D.C., 1993, p. 309.
60. P. Schieberle, New developments in methods for analysis of volatile flavor compounds and their precursors, in *Characterization of Food: Emerging Methods*, A.G. Gaonkar, Ed., Elsevier, New York, 1995, p. 403.
61. A. Kobayashi, Sotolon: identification, formation, and effect on flavor, in *Flavor Chemistry: Trends and Developments*, R. Teranishi, R.G. Buttery, and F. Shahidi, Eds., ACS Symposium Series 388, American Chemical Society, Washington, D.C., 1989, p. 49.
62. W. Grosch, M. Czerny, F. Mayer, and A. Moors, Sensory studies on the key odorants of roasted coffee, in *Caffeinated Beverages: Health Benefits, Physiological Effects, and Chemistry*, T.H. Parliment, C.-T. Ho, and P. Schieberle, Eds., ACS Symposium Series 754, American Chemical Society, Washington, D.C., 2000, pp. 202–209.
63. I. Flament, Coffee, cocoa and tea, in *Volatile Compounds in Foods and Beverages*, H. Maarse, Ed., Marcel Dekker, New York, 1991, p. 617.
64. M. Czerny, F. Mayer, and W. Grosch, Sensory study on the character impact odorants of roasted Arabica coffee, *J. Agric. Food Chem.* 47: 695, 1999.
65. T. Tominaga and D. Dubourdieu, A novel method for quantification of 2-methyl-3-furanthiol and 2-furanmethanethiol in wines made from Vitis vinifera grape varieties, *J. Agric. Food Chem.* 54: 29, 2006.
66. T.H. Parliment, M.J. Morello, and R.J. McGorrin, Eds., *Thermally Generated Flavors*, ACS Symposium Series 543, American Chemical Society, Washington, D.C., 1994.
67. P. Schieberle, Quantitation of important roast-smelling odorants in popcorn by stable isotope dilution assays and model studies on flavor formation during popping, *J. Agric. Food Chem.* 43: 2442, 1995.

68. T. Hofmann and P. Schieberle, 2-Oxopropanal, hydroxyl-2-propanone, and 1-pyrroline — important intermediates in the generation of the roast-smelling food flavor compounds 2-acetyl-1-pyrroline and 2-acetyltetrahydropyridine, *J. Agric. Food Chem.* 46: 2270, 1998.

69. P. Schieberle and W. Grosch, Bread flavor, in *Thermal Generation of Aromas,* T.H. Parliment, R.J. McGorrin, and C.-T. Ho, Eds., ACS Symposium Series 409, American Chemical Society, Washington, D.C., 1989, p. 258.

70. W. Grosch and P. Schieberle, Flavor of cereal products — a review, *Cereal Chem.* 74: 91, 1997.

71. T. Hofmann, R. Hassner, and P. Schieberle, Determination of the chemical structure of the intense roasty, popcorn-like odorant 5-acetyl-2,3-dihydro-1,4-thiazine, *J. Agric. Food Chem.* 43: 2195, 1995.

72. W. Engel and P. Schieberle, Structural determination and odor characterization of N-(2-mercaptoethyl)-1,3-thiazolidine, a new intense popcorn-like-smelling odorant, *J. Agric. Food Chem.* 50: 5391, 2002.

73. R.G. Buttery and L.C. Ling, Importance of 2-aminoacetophenone to the flavor of masa corn flour products, *J. Agric. Food Chem.* 42: 1, 1994.

74. R.G. Buttery and L.C. Ling, Additional studies on flavor components of corn tortilla chips, *J. Agric. Food Chem.* 46: 2764, 1998.

75. M.G. Heydanek and R.J. McGorrin, Gas chromatography-mass spectrometry investigations on the flavor chemistry of oat groats, *J. Agric. Food Chem.* 29: 950, 1981.

76. M.G. Heydanek and R.J. McGorrin, Oat flavor chemistry: principles and prospects, in *Oats: Chemistry and Technology,* F.H. Webster, Ed., *Am. Assoc. Cereal Chem.,* St. Paul, MN, 1986, pp. 335–369.

77. C. Schuh and P. Schieberle, Characterization of (E,E,Z)-2,4,6-nonatrienal as a character impact aroma compound of oat flakes, *J. Agric. Food Chem.* 53: 8699, 2005.

78. G. Zehentbauer and W. Grosch, Crust aroma of baguettes. I. Key odorants of baguettes prepared in two different ways, *J. Cereal Sci.* 28: 81, 1998.

79. T. Hofmann and P. Schieberle, Identification of key aroma compounds generated from cysteine and carbohydrates under roasting conditions, *Z. Lebensm. Unters. Forsch.* 207: 229, 1998.

80. P. Schieberle and P. Pfnuer, Characterization of key odorants in chocolate, in *Flavor Chemistry: Thirty Years of Progress,* R. Teranishi, E.L. Wick, and I. Hornstein, Eds., Kluwer Academic/Plenum, New York, 1999, p. 147.

81. M. van Praag, H.S. Stein, and M.S. Tibetts, Steam volatile constituents of roasted cocoa beans, *J. Agric. Food Chem.* 16: 1005, 1968.

82. G.A. Burdock, Ed., *Fenaroli's Handbook of Flavor Ingredients,* 5th ed., Vol. 2, CRC Press, Boca Raton, FL, 2005.

83. P. Schnermann and P. Schieberle, Evaluation of key odorants in milk chocolate and cocoa mass by aroma extract dilution analyses, *J. Agric. Food Chem.* 45: 867, 1997.

84. J. Maga, Nuts, in *Volatile Compounds in Foods and Beverages,* H. Maarse, Ed., Marcel Dekker, New York, 1991, p. 671.

85. M. Guentert, R. Emberger, R. Hopp M. Koepsel, W. Silberzahn, and P. Werkhoff, Chirospecific analysis in flavor and essential oil chemistry. A. Filbertone-the character impact compound of hazel-nuts, *Z. Lebensm. Unters. Forsch.* 192: 108, 1991.

86. D.A. Withycombe and C.J. Mussinan, Identification of 2-methyl-3-furanthiol in the steam distillate from canned tuna fish, *J. Food Sci.* 53: 658, 1988.

87. D.S. Mottram, Meat, in *Volatile Compounds in Foods and Beverages*, H. Maarse, Ed., Marcel Dekker, New York, 1991, p. 107.

88. C. Cerny and W. Grosch, Quantification of character-impact odor compounds of roasted beef, *Z. Lebensm. Unters. Forsch.* 196: 417, 1993.

89. C. Cerny and W. Grosch, Precursors of ethyldimethylpyrazine isomers and 2,3-diethyl-5-methylpyrazine formed in roasted beef, *Z. Lebensm. Unters. Forsch.* 198: 210, 1994.

90. J. Kerler and W. Grosch, Character impact odorants of boiled chicken: changes during refrigerated storage and reheating, *Z. Lebensm. Unters. Forsch.* 205: 232, 1997.

91. H. Guth and W. Grosch, 12-Methyltridecanal, a species-specific odorant of stewed beef, *Lebensm. Wiss. und Technol.* 26: 171, 1993.

92. J. Kerler and W. Grosch, Odorants contributing to warmed-over flavor (WOF) of refrigerated cooked beef, *J. Food Sci.* 61: 1271, 1996.

93. D.B. Josephson, Seafood, in *Volatile Compounds in Foods and Beverages*, H. Maarse, Ed., Marcel Dekker, New York, 1991, p. 179.

94. F. Shahidi and K.R. Cadwallader, Eds., *Flavor and Lipid Chemistry of Seafoods*, ACS Symposium Series 674, American Chemical Society, Washington, D.C., 1997.

95. R.J. Gordon and D.B. Josephson, Surimi seafood flavors, in *Surimi and Surimi Seafood*, J.W. Park, Ed., Marcel Dekker, New York, 2000 p. 393.

96. C. Milo and W. Grosch, Changes in the odorants of boiled salmon and cod as affected by the storage of the raw material, *J. Agric. Food Chem.* 44: 2366, 1996.

97. C. Milo and W. Grosch, Changes in the odorants of boiled trout (Salmo fario) as affected by the storage of the raw material, *J. Agric. Food Chem.* 41: 2076, 1993.

98. (a). C. Karahadian and R.C. Lindsay, Evaluation of compounds characterizing fishy flavors in fish oils, *J. Am. Oil Chem. Soc.* 66: 953, 1989. (b). A. Kobayashi, K. Kubota, M. Iwamoto, and H. Tamura, Syntheses and sensory characterization of 5,8,11-tetradeca-trien-2-one isomers, *J. Agric. Food Chem.*, 37: 151, 1989.

99. K. Kubota, A. Nakamoto, M. Moriguchi, A. Kobayashi, and H. Ishii, Formation of Pyrrolidino[1,2-e]-4H-2,4-dimethyl-1,3,5-dithiazine in the volatiles of boiled short-necked clam, clam, and corbicula, *J. Agric. Food Chem.* 39: 1127, 1991.

100. F.B. Whitfield, M. Drew, F. Helidoniotis, and D. Svoronos, Distribution of bromophenols in species of marine polychaetes and bryozoans from eastern Australia and the role of such animals in the flavor of edible ocean fish and prawns (shrimp), *J. Agric. Food Chem.* 47: 4756, 1999.

101. L. Schutte, Development and application of dairy flavors, in *Flavor Chemistry: Thirty Years of Progress*, R. Teranishi, E. L. Wick, and I. Hornstein, Eds., Kluwer Academic/Plenum, New York, 1999, p.155.

102. T.H. Parliment and R.J. McGorrin, Critical flavor compounds in dairy products, in *Flavor Chemistry: Industrial and Academic Research*, S.J. Risch and C.-T. Ho, Eds., ACS Symposium Series 756, American Chemical Society, Washington, D.C., 2000, p. 44.

103. R.J. McGorrin, Advances in dairy flavor chemistry, in *Food Flavors and Chemistry: Advances of the New Millennium*, A.M. Spanier, F. Shahidi, T.H. Parliment, C.J. Mussinan, C.-T. Ho, and E.T. Contis, Eds., Royal Society of Chemistry, Cambridge, MA, 2001, pp. 67–84.

104. R. Marsili, Flavors and off-flavors in dairy foods, in *Encyclopedia of Dairy Sciences*, H. Roginski, J. Fuquay, and P.F. Fox, Eds., Elsevier, New York, 2003, pp. 1069–1081.

105. T.K. Singh, M.A. Drake, and K.R. Cadwallader, Flavor of cheddar cheese: a chemical and sensory perspective, *Compr. Rev. Food Sci. Food Saf.* 2: 139, 2003.

106. G. Urbach, The flavor of milk and dairy products: II. Cheese: contribution of volatile compounds, *Int. J. Dairy Technol.* 50: 79, 1997.

107. S. Sable and G. Cottenceau, Current knowledge of soft cheeses flavor and related compounds, *J. Agric. Food Chem.* 47: 4825, 1999.

108. M.E. Carunchia Whetstine, Y. Karagul-Yuceer, Y.K. Avsar, and M.A. Drake, Identification and quantification of character aroma components in fresh chevre-style goat cheese, *J. Food Sci.* 68: 2441, 2003.

109. M. Qian and G.A. Reineccius, Quantification of aroma compounds in Parmigiano Reggiano cheese by a dynamic headspace gas chromatography-mass spectrometry technique and calculation of odor activity value, *J. Dairy Sci.* 86: 770, 2003.

110. M. Qian and G. Reineccius, Static headspace and aroma extract dilution analysis of Parmigiano Reggiano cheese, *J. Food Sci.* 68: 794, 2003.

111. M. Qian and G. Reineccius, Potent aroma compounds in Parmigiano Reggiano cheese studied using a dynamic headspace (purge-trap) method, *Flav. Frag. J.* 18: 252, 2003.

112. C. Milo and G.A. Reineccius, Identification and quantification of potent odorants in regular-fat and low-fat mild cheddar cheese, *J. Agric. Food Chem.* 45: 3590, 1997.

113. G. Zehentbauer and G.A. Reineccius, Determination of key aroma components of Cheddar cheese using dynamic headspace dilution assay, *Flav. Frag. J.* 17: 300, 2002.

114. Y.K. Avsar, Y. Karagul-Yuceer, M.A. Drake, T.K. Singh, Y. Yoon, and K.R. Cadwallader, Characterization of nutty flavor in Cheddar cheese, *J. Dairy. Sci.* 87: 1999, 2004.

115. R. Marsili, Application of SPME GC-MS for flavor analysis of cheese-based products, American Chemical Society, Philadelphia, PA, August 25, 2004.

116. P. Schieberle, K. Gassenmeier, H. Guth, A. Sen, and W. Grosch, Character impact odor compounds of different kinds of butter, *Lebensm. Wiss. und Technol.* 26: 347, 1993.

117. Y. Karagul-Yuceer, M.A. Drake, and K.R. Cadwallader, Aroma-active components of nonfat dry milk, *J. Agric. Food Chem.* 49: 2948, 2001.

118. S.S. Mahajan, L. Goddik, and M.C. Qian, Aroma compounds in sweet whey powder, *J. Dairy Sci.* 87: 4057, 2004.

119. M.E. Carunchia Whetstine, A.E. Croissant, and M.A. Drake, Characterization of dried whey protein concentrate and isolate flavor, *J. Dairy Sci.* 88: 3826, 2005.

120. J.T. Budin, C. Milo, and G.A. Reineccius, Perceivable odorants in fresh and heated sweet cream butters, in *Food Flavors and Chemistry: Advances of the New Millennium*, A.M. Spanier, F. Shahidi, T.H. Parliment, C.J. Mussinan, C.-T. Ho, and E.T. Contis, Eds., Royal Society of Chemistry, Cambridge, MA, 2001, pp. 85–96.

121. (a) D.G. Peterson and G.A. Reineccius, Characterization of the volatile components which constitute fresh sweet cream butter aroma, *Flav. Frag. J.* 18: 215, 2003. (b) D.G. Peterson and G.A. Reineccius, Determination of the aroma impact compounds in heated sweet cream butter, *Flav. Frag. J.* 18: 320, 2003.

122. M. Preininger and W. Grosch, Evaluation of key odorants of the neutral volatiles of Emmentaler cheese by the calculation of odor activity values, *Lebensm. Wiss. und Technol.* 27: 237, 1994.

123. A. Ott, L.B. Fay, and A. Chaintreau, Determination and origin of the aroma impact compounds of yogurt flavor, *J. Agric. Food Chem.*, 45: 850, 1997.

124. M.J. Saxby, *Food Taints and Off-Flavors*, 2nd ed, Blackie Academic and Professional, London, 1996, p. 326.

125. G. Reineccius, Off-flavors and taints in foods, in *Flavor Chemistry and Technology*, 2nd ed., CRC Press/Taylor and Francis, Boca Raton, FL, 2006, pp. 161–200.

126. B. Nijssen, Off-flavors, in *Volatile Compounds in Food and Beverages*, H. Maarse, Ed., Marcel Dekker, New York, 1991, p. 689.

127. G. Charalambous, Ed., *Off-Flavors in Foods and Beverages*, Elsevier, New York, 1992.

128. R. Marsili, Off-flavors and malodors in foods: mechanisms of formation and analytical techniques, in *Techniques for Analyzing Food Aroma*, R. Marsili, Ed., Marcel Dekker, New York, 1997, p. 237.

129. S. Widder and W. Grosch, Precursors of 2-nonenals causing the cardboard off-flavor in butter oil, *Nahrung* 41: 42, 1997.

130. (a). A. Stephan and H. Steinhart, Identification of character impact odorants of different soybean lecithins, *J. Agric. Food Chem.* 47: 2854, 1999. (b). W.L. Boatright and Q. Lei, Compounds contributing to the beany odor of aqueous solutions of soy protein isolates, *J. Food Sci.* 64: 667, 1999.

131. (a) H. Guth and W. Grosch, 3-Methylnonane-2,4-dione — an intense odor compound formed during flavor reversion of soybean oil, *Fat Sci. Technol.* 91: 225, 1989. (b) H. Guth and W. Grosch, Impact of 3-methylnonane-2,4-dione on the flavor of oxidized soybean oil, *J. Am. Oil Chem. Soc.* 76: 145, 1999.

132. W.E. Neff and E. Selke, Volatile compounds from the triacylglycerol of cis,cis 9,15-linoleic acid, *J. Am. Oil Chem. Soc.* 70: 157, 1993.

133. S. Noel, C. Liegeois, G. Lermusieau, E. Bodart, C. Badot, and S. Collin, Release of deuterated nonenal during beer aging from labeled precursors synthesized in the boiling kettle, *J. Agric. Food Chem.* 47: 4323, 1999.

134. K. Harayama, F. Hayase, and H. Kato, Contribution to stale flavor of 2-furfuryl ethyl ether and its formation mechanism in beer, *Biosci. Biotechnol. Biochem.* 59: 1144, 1995.

135. K. Jensen, M.A. Petersen, L. Poll, and P.B. Brockhoff, Influence of variety and growing location on the development of off-flavor in precooked vacuum-packed potatoes, *J. Agric. Food Chem.* 47: 1145, 1999.

136. C. Masanetz, H. Guth, and W. Grosch, Fishy and hay-like off-flavors of dry spinach, *Z. Lebensm. Unters. Forsch.* 206: 108, 1998.

137. C. Masanetz, H. Guth, and W. Grosch, Hay-like off-flavor of dry parsley, *Z. Lebensm. Unters. Forsch.* 206: 114, 1998.

138. M. Kamimura and H. Kaneda, Off-flavors in beer, in *Off-Flavors in Foods and Beverages*, G. Charalambous, Ed., Elsevier, New York, 1992, p. 433.

139. S. Masuda, K. Kikuchi, K. Harayama, K. Sakai, and M. Ikeda, Determination of lightstruck character in beer by gas chromatography-mass spectrometry, *J. Am. Soc. Brew. Chem.* 58: 152, 2000.

140. J. Templar, K. Arrigan, and W.J. Simpson, Formation, measurement and significance of lightstruck flavor in beer: a review, *Brewers' Digest* 70: 18, 1995.

141. J.F. Clapperton, Ribes flavor in beer, *J. Inst. Brew.* 82: 175, 1976.

142. K.B. Cosser, J.P. Murray, and C.W. Holzapfel, Investigation of a ribes off-flavor in beer, *Tech. Q. Master Brew. Assoc. Am.* 17: 53, 1980.

143. D.A. Escudero, P. Hernandez-Orte, J. Cacho, and V. Ferreira, Clues about the role of methional as character impact odorant of some oxidized wines, *J. Agric. Food Chem.* 48: 4268, 2000.

144. P. Perpete and S. Collin, Contribution of 3-methylthiopropionaldehyde to the worty flavor of alcohol-free beers, *J. Agric. Food Chem.* 47: 2374, 1999.

145. H.-D. Belitz and W. Grosch, *Food Chemistry,* 2nd English ed., Springer-Verlag, Berlin, 1999, p. 322.

146. R.L. Swaine, T.S. Myers, C.M. Fischer, N. Meyer, and B. Pohlkamp, The formation of off-flavors in fortified juice-containing beverages: identification of 4-methyl-2-isopropylthiazole, *Perf. Flav.,* 22: 57, 1997.

147. K. Bernath, Sulfurous off-flavours in wine. [Das Boeckser-aroma in Wein], *Diss. Abstr. Int. C,* 59: 308, 1998.

148. A. Anocibar-Beloqui, P. Guedes de Pinho, and A. Bertrand, Bis(2-hydroxyethyl) disulfide, a new sulfur compound found in wine: its influence in wine aroma, *Am. J. Enol. Viticult.* 46: 84, 1995.

149. J.L. Licker, T.E. Acree, and T. Henick-Kling, What is "Brett" (*Brettanomyces*) flavor?: a preliminary investigation, in *Chemistry of Wine Flavor,* A.L. Waterhouse and S.E. Ebeler, Eds., ACS Symposium Series 714, American Chemical Society, Washington, D.C., 1998, pp. 96–115.

150. E. Chambers IV, E.C. Smith, L.M. Seitz, and D.B. Sauer, Sensory properties of musty compounds in food, in *Food Flavors: Formation, Analysis, and Packaging Influences,* E.T. Contis, C.-T. Ho, C.J. Mussinan, T.H. Parliment, F. Shahidi, and A.M. Spanier, Eds., Elsevier, New York, Developments in Food Science Vol. 40, 1998, p. 173.

151. L.H. Aung, J.L. Smilanick, P.V. Vail, P.L. Hartsell, and E. Gomez, Investigations into the origin of chloroanisoles causing musty off-flavor of raisins, *J. Agric. Food Chem.* 44: 3294, 1996.

152. J.C. Spadone, G. Takeoka, and R. Liardon, Analytical investigation of Rio off-flavor in green coffee, *J. Agric. Food Chem.* 38: 226, 1990.

153. O.G. Vitzthum, Thirty years of coffee chemistry research, in *Flavor Chemistry: Thirty Years of Progress,* R. Teranishi, E.L. Wick, and I. Hornstein, Eds., Kluwer Academic/Plenum, New York, 1999, p. 117.

154. S.W. Lloyd and C.C. Grimm, Analysis of 2-methylisoborneol and geosmin in catfish by microwave distillation — solid phase microextraction, *J. Agric. Food Chem.* 47: 164, 1999.

155. Y. Jirawan and N. Athapol, Geosmin and off-flavor in Nile tilapia (*Oreochromis niloticus*), *J. Aquat. Food Prod. Technol.* 9: 29, 2000.

156. M.R. Sevenants and R.A. Sanders, Anatomy of an off-flavor investigation: the "medicinal" cake mix, *Anal. Chem.* 56: 293A, 1984.

157. R.J. McGorrin, T.R. Pofahl, and W.R. Croasmun, Identification of the musty component from an off-odor packaging film, *Anal. Chem.* 59: 1109A, 1987.

158. R.A. Sanders, D.V. Zyzak, T.R. Morsch, S.P. Zimmerman, P.M. Searles, M.A. Strothers, B.L. Eberhart, and A.K. Woo, Identification of 8-nonenal as an important contributor to "plastic" off-odor in polyethylene packaging, *J. Agric. Food Chem.* 53: 1713, 2005.

159. O.W. Parks, D.P. Schwartz, and M. Keeney, Identification of o-aminoacetophenone as a flavor compound in stale dry milk, *Nature* 202: 185, 1964.

160. E.H. Ramshaw and E.A. Dunstone, Volatile compounds associated with the off-flavor in stored casein, *J. Dairy Res.* 36: 215, 1969.
161. B. Dollmann, A. Schmitt, H. Koehler, and P. Schreier, [Formation of the "untypical ageing off-flavor" in wine: generation of 2-aminoacetophenone in model studies with *Saccharomyces cerevisiae*], *Wein-Wiss.* 51: 122, 1996.
162. W. Stark and D.A. Forss, A compound responsible for metallic flavor in dairy products. I. Isolation and identification, *J. Dairy Res.* 29: 173, 1962.
163. M.B. Lubran, H.T. Lawless, E. Lavin, and T.E. Acree, Identification of me-tallic-smelling 1-octen-3-one and 1-nonen-3-one from solutions of ferrous sulfate, *J. Agric. Food Chem.* 53: 8325, 2005.
164. M.Y. Jung, S.H. Yoon, H.O. Lee, and D.B. Min, Singlet oxygen and ascorbic acid effects on dimethyl disulfide and off-flavor in skim milk exposed to light, *J. Food Sci.* 63: 408, 1998.
165. T. Koenig, B. Gutsche, M. Hartl, R. Huebscher, P. Schreier, and W. Schwab, 3-Hydroxy-4,5-dimethyl-2(5H)-furanone (sotolon) causing an off-flavor: elucidation of its formation pathways during storage of citrus soft drinks, *J. Agric. Food Chem.* 47: 3288, 1999.
166. R.G. Buttery, W.F. Haddon, R.M. Seifert, and J.G. Turnbaugh, Thiamin odor and bis(2-methyl-3-furyl)disulfide, *J. Agric. Food Chem.* 32: 674, 1984.
167. C.J. Scarlata and S.E. Ebeler, Headspace solid phase microextraction for the analysis of dimethyl sulfide in beer, *J. Agric. Food Chem.* 47: 2505, 1999.
168. C. Heiler and P. Schieberle, Model studies on the precursors and formation of the metallic smelling (E,Z)-2,6-nonadienol during the manufacture and storage of buttermilk, *Int. Dairy J.* 7: 667, 1997.
169. A. Sensidoni, G. Rondinini, D. Peressini, M. Maifreni, and R. Bortolomeazzi, Presence of an off-flavor associated with the use of sorbates in cheese and margarine, *Ital. J. Food Sci.* 6: 237, 1994.
170. H. Shiratsuchi, Y. Yoshimura, M. Shimoda, K. Noda, and Y. Osajima, Objective evaluation of off-flavor in spray-dried skim milk powder. *J. Jpn. Soc. Food Sci. Technol.* 43: 7, 1996.
171. M. Preininger and F. Ullrich, Trace compound analysis for off-flavor charac-terization of micromilled milk powder, in *Gas Chromatography-Olfactometry: The State of the Art*, J.V. Leland, P. Schieberle, A. Buettner, and T.E. Acree, Eds., ACS Symposium Series 782, American Chemical Society, Washington, D.C., 2001, p. 46.
172. Y. Karagul-Yuceer, K.N. Vlahovich, M.A. Drake, and K.R. Cadwallader, Char-acteristic aroma components of rennet casein, *J. Agric. Food Chem.* 51: 6797, 2003.
173. K. Iwatsuki, Y. Mizota, T. Kubota, O. Nishimura, H. Matsuda, K. Sotoyama, and M. Tomita, Evaluation of aroma of pasteurized and UHT processed milk by aroma extract dilution analysis, *J. Jpn. Soc. Food Sci.* 46: 587, 1999.
174. P.A. Vazquez-Landaverde, G. Velazquez, J.A. Torres, and M.C. Qian, Quantita-tive determination of thermally derived off-flavor compounds in milk using solid-phase microextraction and gas chromatography, *J. Dairy Sci.* 88: 3764, 2005.
175. P.M. Colahan-Sederstrom and D.G. Peterson, Inhibition of key aroma com-pound generated during ultrahigh-temperature processing of bovine milk during epicatechin addition, *J. Agric. Food Chem.*, 53: 398, 2005.
176. F.B. Whitfield, N. Jensen, and K.J. Shaw, Role of *Yersinia intermedia* and *Pseudomonas putida* in the development of a fruity off-flavor in pasteurized milk, *J. Dairy Res.* 67: 561, 2000.

177. D.A. Zabolotsky, L.F. Chen, J.A. Patterson, J.C. Forrest, H.M. Lin, and A.L. Grant, Supercritical carbon dioxide extraction of androstenone and skatole from pork fat, *J. Food Sci.* 60: 1006, 1995.
178. Y. Tanaka, T. Sasao, T. Kirigaya, S. Hosoi, A. Mizuno, T. Kawamura, and H. Nakazawa, Analysis of skatole in off-flavor beef by GC/MS, *J. Food Hyg. Soc. Jpn.* 39: 281, 1998.
179. A.M. Popken, H.M. Dechent, and D. Guerster, Investigations on the origin of carvone in orange juices as an off-flavor component, *Fruit Process* 9: 338, 1999.
180. M. Walsh, R. Rouseff, and M. Naim, Determination of furaneol and p-vinylguaiacol in orange juice employing differential UV wavelength and fluorescence detection with a unified solid phase extraction, *J. Agric. Food Chem.* 45: 1320, 1997.
181. J.A. Donaghy, P.F. Kelly, and A. McKay, Conversion of ferulic acid to 4-vinyl guaiacol by yeasts isolated from unpasteurised apple juice, *J. Sci. Food Agric.* 79: 453, 1999.
182. D. Madigan, I. McMurrough, and M.R. Smyth, Rapid determination of 4-vinyl guaiacol and ferulic acid in beers and worts by high-performance liquid chromatography, *J. Am. Soc. Brew. Chem.* 52: 152, 1994.

Index

S